Food and Energy

DATE DUE

MY23'00			

DEMCO 38-296

Note to the Reader from the UNU

The Food-Energy Nexus Programme was launched in 1983 to fill the research gap that existed on the synergistic solutions to food and energy problems. The programme consisted of a two-pronged effort directed towards developing an analytical framework and planning methodology as well as stimulating the sharing of experiences between research teams working in Asia, Africa, and Latin America. It addressed such concrete issues as (i) a more efficient use of energy in the production, processing, and consumption of food; (ii) food-energy systems in diverse ecosystems; and (iii) household economy in both rural and urban settings and the role of women and children in the provision of food, fuel, and water. *Food and Energy: Strategies for Sustainable Development* sums up the research findings and their policy implications in comparative regional perspectives.

R

Food and Energy
Strategies for Sustainable Development

Ignacy Sachs and Dana Silk

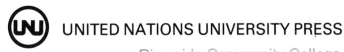 UNITED NATIONS UNIVERSITY PRESS

The views expressed in this publication are those of the authors and do not necessarily reflect the views of the United Nations University.

United Nations University Press
The United Nations University, Toho Seimei Building, 15-1 Shibuya 2-chome, Shibuya-ku, Tokyo 150, Japan
Tel.: (03) 499-2811   Fax: (03) 499-2828
Telex: J25442   Cable: UNATUNIV Tokyo

Typeset by Asco Trade Typesetting Limited, Hong Kong
Printed by Permanent Typesetting and Printing Co. Ltd., Hong Kong
Cover design by Tsuneo Taniuchi

FEN-1/UNUP-757
ISBN 92-808-0757-9
United Nations Sales No. E.90.III.A.12
02500 P

Contents

Preface

This publication is based on the activities of the Food-Energy Nexus Programme (FEN) of the United Nations University (UNU) which took place between 1983 and 1988. As is the custom with the UNU, most of this research was done by researchers associated with various universities or research centres around the world. In this case, considerable work was done by researchers in third world countries in order to promote South-South co-operation in the fields studied.

We wish to express our gratitude to the many people who contributed in various ways to FEN and to acknowledge the institutional support of the Maison des Sciences de l'Homme in Paris. At the same time, we assume entire responsibility for any misinterpretations or omissions that may be apparent in the following chapters. For more detailed information, readers are referred to Appendices I and II, which reprint the descriptions of FEN programme activities and publications found in its final report.

We hope that this complementary analysis of FEN concepts will contribute to the preparatory process for the second United Nations Conference on the Environment and Development in 1992 by helping to identify future research, development, and implementation activities for the sustainable production of, and more equitable access to, the basic human needs of food and energy.

1

Introduction

For most people, the relation between food and energy problems first became evident as a result of the oil crisis in the early 1970s. While immediate attention was given by industrialized countries to ensuring adequate oil supplies to fuel their energy-intensive food systems, long-term concerns were raised about the plight of the rural and urban poor in third world countries with the realization that the high cost of energy and fertilizers would further limit the scope of the Green Revolution.

Beyond the oil price problem loomed the second energy crisis, with even greater social and ecological consequences for more than half of the world's population. In practically all third world countries the problems of getting food to eat began to be overshadowed by the problems of acquiring the energy needed to cook it. Apart from the financial sacrifices, there was a severe strain on time budgets, notably those of women and children, who spend increasingly long hours collecting fuelwood (Cecelski 1987). These problems are exacerbated by the seasonal imbalance in biomass supply and the vicious cycle of greater quantities of dung being used as fuel rather than as fertilizer for maintaining crop production.

These developments have already been amply documented, but much less research is available on the synergistic solutions to food and energy problems. The Food-Energy Nexus Programme (FEN) of the United Nations University (UNU) was thus created to help fill this gap through a two-pronged effort: to develop an analytical framework and planning methodology; and to stimulate the sharing of experiences between research teams working in Asia, Africa, and Latin America.

Speaking at an intergovernmental meeting of development assistance co-ordinators in Asia and the Pacific, held in February 1981 in New Delhi, the Rector of the UNU, Soedjatmoko (1981), first made reference to what was to become FEN:

Rising fuel prices, boosting transportation and agricultural costs, will inevitably push food prices beyond the reach of hundreds of millions of already hungry people. Rising populations, despite the best efforts to reduce fertility rates, will continue to increase the demand for both food and energy. The developing countries will not be able to solve their food problem without solving their energy problem and, without a satisfactory solution to both, their economic growth will be severely constrained. The centrality of the food and energy nexus calls for a comprehensive policy approach. Only through a clear understanding of this food-energy pivot can the situation be turned around.

The following year, Sachs (1982) described the food-energy nexus as a convenient entry point into the problematique of efficient resource use patterns for sustainable development and of local solutions to global problems.

FEN was thus initially designed to develop the following activities:
- policy-oriented studies of the energy profile of food systems to identify options leading to a more efficient use of energy in the production, processing, distribution, preparation, and consumption of food;
- analyses of integrated food-energy systems in diverse ecosystems, including studies of how different cultures manage to extract from similar environments their food and energy supplies; and
- research on the household economy in both rural and urban settings and the role of women and children in the provisioning of food, fuel, and water.

Further attention to the conceptual development of FEN resulted in the sharpening of its focus around two axes: integrated food-energy systems as a catalyst for rural development and industrialization; and alternative urban development strategies based on greater self-reliance. The FEN description that was prepared in 1985 remained more or less unchanged during the lifespan of the programme:

FEN is predicated on the idea that positive synergies can be developed by addressing simultaneously the ideas of production and access to food and fuel and by building around these twin objectives self-reliant development strategies. It thus promotes action-oriented, interdisciplinary research in the following four areas:
1. integrated food-energy systems based on closed-loop ecological models and adapted to site-specific environmental and cultural conditions;
2. social innovations in the urban setting leading to greater equity, efficiency, and sustainability in the use of resources to improve access to food and energy by the urban poor;
3. social impacts of food processing and energy-producing technologies; and
4. adaptation of resource-use patterns in diverse ecosystems for the provisioning of food and energy to the rural and urban poor.

The broad scope of FEN and the modest resources that it had available pointed naturally to the significance of its "enzyme" role. It was thus active in inspiring projects funded from local sources, assisting third-world-based research organizations in the design of such projects, stimulating real-size experiments, collecting and disseminating information in the form of state-of-the-art reports, and promoting the exchange of scholars and bringing them together in workshops and through networks. Special attention was paid to the "third system" or citizens' organizations through, *inter alia*, the urban self-reliance project that was implemented with the International Foundation for Development Alternatives (IFDA) in Nyon, Switzerland.

From the very beginning, an effort was made to give a global learning dimension to FEN and to emphasize South-South co-operation. This began with a study tour by four Brazilian researchers to Senegal, India, and China and was facilitated by FEN conferences that were subsequently held in Latin America, Asia, and Africa. This policy of strengthening research capacities in third world countries and improving communication between people working on similar issues was also adopted by FEN projects in the urban field.

2

Analytical Framework

Local Solutions to Global Problems

Before referring to specific work on integrated food-energy systems in rural areas and alternative urban development strategies, it may be useful to highlight the conceptual framework which provided the intellectual impetus for FEN.

Food and energy are global problems in at least four ways. First, their regular and continuing availability is a condition *sine qua non* of human survival, posing a formidable challenge that must be tackled simultaneously from both the supply and demand sides. Food and fuel stocks are of no help and little consolation to people who cannot afford to buy them and have no access to the resources needed to produce them.

Second, assuming optimistically that humanity will manage to solve the problems posed by its bare survival, the quality of life of millions of people will still depend to a great extent on increased supplies and better use of both food and energy; their central role in a need-oriented development strategy is only too obvious.

Third, both food and energy loom large in the North-South confrontation as potential weapons and tools of domination; hence the importance of global negotiations to modify the present gloomy picture and bring about some constructive international co-operation in both fields.

Finally, food and energy production affect and are affected by the state of the environment: energy-, land-, and water-use patterns will increasingly influence the climate and other aspects of our life-support systems. This will have far-reaching consequences for the long-term prospects of food and energy production in semi-arid areas and threatens the very existence of flood-prone coastal settlements, home to millions of people.

In contrast with the considerable research effort that has been spent on each of these problems in their own right, however, little attention has been paid to the food-energy nexus or the systematic exploration of the ways in which they

are linked. Must we be reminded that the prospects for both of these resources, treated individually, are rather dismal?

According to the *Global 2000 Report* (Council on Environmental Quality
1980), the amount of arable land per person is projected to decrease from
about 0.4 ha in 1975 to about 0.25 ha by the end of the century. If current
trends continue, the world's per capita growing stock of wood will be 47 per
cent lower in the year 2000 than in 1978 and 40 per cent of the forests still
remaining in the third world will have been razed. Under such conditions, the
real price of food is likely to multiply, along with long-term declines in the productivity of such over-taxed renewable resource systems.

Furthermore, according to a study by the International Institute for Applied
Systems Analysis, 30 to 70 per cent of the intermediate input costs of agricultural crop production in developing countries are directly or indirectly related to
energy. On the other hand, agriculture provides 20 to 90 per cent of primary
energy through the supply of the so-called non-commercial energy sources
such as wood and agricultural residues (Parikh 1981).

The intellectual challenge is to go beyond the simple analysis of data and
assist governments in transcending the current crisis management approach
by embracing development planning conceived as a societal learning process.
This calls for the enhancement of the capacity of social organizations and individuals to respond creatively to new situations, constraints, and opportunities
and to shape new trajectories for industrialization and rural modernization that
are different from the models derived from the historical experiences of industrialized countries of the North.

Projections serving in lieu of forecasts can only lead to mistaken conclusions
when they concentrate on quantitative growth within ossified structures rather
than anticipating structural changes and modifications in social behaviour. Specific local solutions responding to the variety of ecological, cultural, socio-
economic, and political contexts in today's world must thus be found to the
global problems of food and energy.

The *differentia specifica* of FEN consisted in its concentration on planning,
designing, implementing, and evaluating ecosystem-specific, culture-specific,
and even site-specific solutions to global problems. Such research must necessarily be done downstream from other work on food and energy macro-models,
such as the Global Food-Energy Modelling Project (Robinson 1986) that FEN
was associated with.

Useful as the latter may be to sensitize policy-makers about the long-term
implications of their decisions, or to test the coherence of proposed policies,
they do not lend themselves to straightforward translation into planning guide-
lines at regional, micro-regional, or project level. Macro-models must perforce
assume away the diversity of local conditions and thus cannot do justice to people's resourcefulness, a key concept for successful ecodevelopment planning.

Entitlement and Ecodevelopment

FEN was premised on the "entitlement" approach articulated by Amartya Sen and on the concept of ecodevelopment. In a recent publication, Sen (1987) aptly summarizes the entitlement approach to food, which must be extended equally to energy for cooking:

The real issue is not primarily the overall availability of food, but its acquirement by individuals and families. If a person lacks the means to acquire food, the presence of food in the market is not much consolation. To understand hunger, we have to look at people's entitlement, i.e. what commodity bundles (including food) they can make their own. The entitlement approach to hunger concentrates on the determination of command over commodities, including food. Famines are seen as the result of entitlement failures of large groups, often belonging to some specific occupations (e.g. landless rural labourers, pastoralists).

The entitlement of a person stands for the set of different alternative commodity bundles that the person can acquire through the use of the various legal channels of acquirement open to someone in his position. In a private ownership market economy, the entitlement set of a person is determined by his original bundles of ownership (what is called his "endowment") and the various alternative bundles he can acquire starting from each initial endowment, through the use of trade and production (what is called his "exchange entitlement mapping"). A person has to starve if his entitlement set does not include any commodity bundles with adequate amounts of food. A person is reduced to starvation if some change either in his endowment (e.g. alienation of land, or loss of labour power due to ill health) or in his exchange entitlement mapping (e.g. fall in wages, rise in food prices, loss of employment, drop in the price of the good he produces and sells), makes it no longer possible for him to acquire any commodity bundle with enough food.

As for ecodevelopment (Sachs 1980), it is based on the simultaneous pursuit of the following four objectives:
1. *Social equity*, that is, better access by low-income people to goods and services needed for a decent life. This can be achieved through a variety of means deemed culturally desirable by those concerned: employment and higher incomes, generation of opportunities for self-production and self-help construction (assisted by the state), subsidized housing, free education and health services, and, in the case of poverty, food distribution through special programmes.
2. *Ecological sustainability* both in terms of resource conservation and the minimization of the harmful impacts of production on the environment and, by extension, on people's health and quality of life.
3. *Economic efficiency* considered at the macro-social level, that is, ensuring a rational pattern of resource use for the whole society, the aim of the economic policy being to create such conditions for private enterprises as to

make the micro-entrepreneurial criteria coincide to a large degree with the social ones.
4. *Balanced spatial distribution* of human settlements and activities so as to avoid some of the worst problems resulting from excessive concentration or dispersion of human endeavours.

The concept of ecodevelopment emphasizes the sustainable use of local human and natural resources for meeting locally defined needs. It therefore embraces a radically participatory approach. People's needs must be defined realistically and autonomously so as to avoid the harmful "demonstration effects" of the consumption style of rich countries. Since people are the most valuable resource, ecodevelopment should contribute primarily to their self-realization.

This calls for the establishment of a horizontal authority which can transcend sectoral particularism, an authority concerned with all the facets of development and which makes continual use of the complementarity of the different actions undertaken. To be efficient, such an authority requires the effective participation of the people concerned.

In other words, ecodevelopment is closely related to the "strong" version of the basic needs approach defined by Wisner (1988) in contrast with the "weak" version. The former encourages poor people to understand the social origins of their poverty and to struggle to change them. Insofar as the latter involves the delivery of a bundle of goods and services, it treats the poor as passive recipients, not activists capable of helping themselves.

The Urban Challenge

If current urbanization trends continue, by the year 2010 – just one generation from now – low-income people in third world cities will become the new majority among the world's population, displacing the rural poor.

All three developing continents are facing very serious urban problems. The backlog of unattended needs for housing, food, energy, and services is so great and the pace of urbanization so rapid that it has become physically and financially impossible to solve this situation by providing more of the same. The conventional solutions, which proved more or less effective in industrialized countries, simply cannot be afforded unless there is a drastic reallocation of world resources, currently directed at military spending. This sane alternative to the arms race, however, appears highly unlikely.

Thus, the degree of satisfaction of the basic needs of the growing urban populations will, to a great extent, depend on the creativity and resourcefulness of the communities themselves as well as of their administrators. New urban development strategies are called for, based on social innovations.

Let us start by enumerating the five main fields of potential innovation.

The first deals with new forms of organization of economic activity, capable of improving the degree of utilization of human potential for work available in society. In order to ascertain it, we must better understand the everyday structures of material life, the cultural models of time use, and the working of the "real economy". This term encompasses the complex web of interconnected markets of labour, goods, and services ranging from the official to the criminal, as well as the non-market household economy, the embryonic forms of the non-market social economy, and the multiple interventions of the state. In other words, it is necessary to go beyond the formal/informal dichotomy and to analyse both the monetary flows and the cultural patterns of time allocation for work in the market-oriented and non-market economic endeavours, as well as for non-economic activities (Sanchez 1988).

The second consists of the untapped, under-utilized, misused, or wasted resources existing in the urban ecosystem: vacant land, waste, and sewage that can be recycled or reused, energy and water that can be conserved at a lower cost than the production of additional supplies, etc. Such resources, detected through an ecological analysis, offer interesting opportunities for employment and/or self-employment, often requiring only a moderate investment per worker. More generally, countries short of capital ought to pay the utmost attention to "non-investment" sources of growth, such as maintenance of existing equipment, elimination of wasteful resource-use patterns, recycling, etc.

The third sphere encompasses the whole area of identification and production of appropriate technologies, enabling a rational and more intensive use of the capacity to work of people and available physical resources. The concept of appropriate technology is, once more, relative to a given ecological, cultural, and socio-economic context. There exist no appropriate technologies, as such, that are universally applicable. Nor is it reasonable to choose, *a priori*, capital-intensive technologies or labour-intensive ones. A selective use of the whole range of technologies is called for and scientists should be encouraged to search for new "knowledge-intensive", resource-conserving technologies.

The fourth domain is the institutional one, perhaps the most difficult and, in many respects, the most decisive. Development cannot be left to market forces alone, nor be made an exclusive responsibility of the state. Community involvement is indispensable at all levels: setting of priorities, creativity in searching for solutions, participation in implementing them. Institutional breakthroughs will only happen through new forms of partnership for development between society, the market, and the state. The whole field of party politics, non-party politics related to citizens' organizations, and workers' self-management within enterprises needs to be reconceptualized.

Finally, *the fifth* field for innovations in the urban setting is concerned with public policy instruments and the selection of policy packages that stimulate and support social innovations. The emphasis on community participation and self-help programmes should not be taken as a pretext to reduce the responsibility and the share of the state in carrying out vigorous activities aimed at alleviating the plight of low-income urban populations.

In the following chapters, some of the ways in which this conceptual approach was exploited by FEN researchers are described. It should be borne in mind that the interdisciplinary and evolutive nature of this approach was consciously designed to stimulate work in this field, not to confine it to a preconceived framework.

3

Integrated Food-Energy Systems

One of the two major focuses of FEN was the analysis of systems designed to integrate, intensify, and thus increase the production of food and energy by transforming the by-products of one system into the feedstocks for the other. Such integrated food-energy systems (IFES) can operate at various scales, ranging from the industrial-sized operations in Brazil designed to produce primarily ethanol and fertilizer (La Rovere and Tolmasquim 1986), to the village- and even household-level biogas systems in India (Moulik 1985). A conceptual outline of such systems, including the long-ignored but vitally important component of water for biomass-based production, is shown in figure 1.

This valorization of agricultural by-products plays an important role in the design of IFES for specific agro-climatic regions, designs which should closely follow the paradigm of natural ecosystems. These modern, ecologically sound systems are characterized by their closed loops of resource flows. In a sense, they incorporate the same rationality of traditional peasant farmers, albeit at a completely different level of scientific and technical knowledge.

The study of such systems, adapted to the diversity of natural environments and responding to a wide spectrum of needs in terms of size, technological sophistication, and capital intensity, became the main thrust of FEN research activities in this field. Earlier UNU work on patterns of resource use and management in Asian villages (Ruddle and Manshard 1981) provided a very useful starting point for this research.

Two international conferences in this field were held in Brasilia (FINEP/ UNESCO 1986) and New Delhi (Moulik 1988), organized with the co-operation of UNESCO as well as that of Brazilian and Indian authorities, research organizations, and universities. They provided a valuable forum to review on-going work on IFES and to discuss future research priorities (Wisner 1986). These conferences were complemented by a subsequent symposium in Changzhou, organized by the Chinese Association for Science and Technology, which compared IFES in seven ecologically diverse regions of China.

Such events were instrumental in showing that this concept had attracted considerable attention in many developing countries, leading to both research

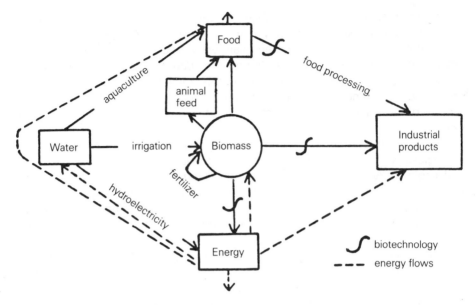

Fig. 1. A systems approach to rural development

and practical implementation. These meetings were also very important in fostering South-South co-operation between researchers and government officials in the countries involved.

Integrated Systems

For our purposes, the word "integrated" has three superimposed meanings: multiple sources of inputs (including energy), multi-task devices, and time sharing of devices (such as engines). Techniques of different vintages are also used in an attempt to accommodate heterogeneity rather than to impose a homogenizing modernity across the board resulting in a massive displacement of labour.

The design of IFES requires a simultaneous consideration of the biophysical components of resource management, of the social and ecological impacts of the technologies used, and of the institutional settings involved. For each site-specific configuration of climatic and environmental conditions, several socially desirable, ecologically sustainable, and economically efficient production systems are conceivable, differing in output mix, forms of social organization and community participation, size of operation, complexity of design, and technical sophistication. Ideally, they should have a modular structure, allowing for a progressive implementation by adding new modules to the initial structure.

By comparing several systems, observing their performance, and exchanging experiences and results between projects situated in similar ecosystems and different cultural areas, development planners would come as near to a social laboratory setting as it is possible. In order to adapt proposed solutions to the specific local conditions and needs, community involvement is required at all stages: to identify not only the pressing problems but also the latent resources, then to progressively build the system, and finally to manage it in a proactive way.

Special attention should also be given to integrated systems designed for ecologically vulnerable areas and for reclaiming wastelands. Socially responsive and ecologically sustainable agroforestry systems, protecting both the local people and the trees, should receive maximum attention. Agroforestry is possible even under adverse climatic conditions, although the repeated failures of ambitious schemes in the Amazon region should teach us modesty and patience.

As for wastelands, considerable scope exists for establishing fish farms as part of mixed farming systems. In many maritime states of India, for example, large areas of saline soils, marshy swamps, and mangroves are lying fallow. Their reclamation for agriculture may be costly but such resources could be used for brackish water aquaculture generating both food and work for landless peasants.

Ecosystems, Food, and Energy[1]

The 1984 Brasilia seminar described IFES as the adoption of agricultural and industrial technologies that allow maximum utilization of by-products, diversification of raw materials, production on a small-scale, recycling and economic utilization of residues, and harmonization of energy and food production. Such systems imply the need for comprehensive land-use planning as well as planned interrelations among soil, water, and forest resources in relation to agricultural residues. Major advantages are their minimal negative environmental impact, and their decentralization and efficiency, which often have positive social and economic side-effects.

In Brazil, IFES notably include micro-distilleries for alcohol production from sugar cane, sweet sorghum, manioc or sugar beets; fattening of stall-fed livestock; biodigestors for decomposition of livestock manure and/or sugar cane bagasse and/or stillage; generation of electricity and agricultural mechanization on the basis of the fuels thus produced; and the application of biofertilizer to cultivated lands (fig. 2). La Rovere and Tolmasquim (1986) note the possibility of extending this to include various ways of enhancing the water management system, including the production of aquatic plants, fish, and zooplankton as well as the retention of water for sedimentation and irrigation purposes.

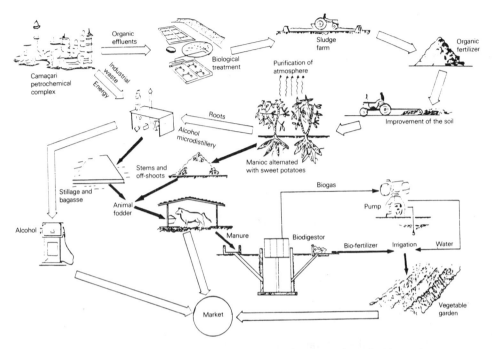

Fig. 2. Flowchart of the "pronatura" food-energy integrated production scheme proposed for the petrochemical complex of Camaçari, Bahia (reproduced from the *Food-energy Nexus Newsletter* 2, no. 2 [September 1985])

As is noted above, IFES can be put together in various configurations and at various scales. A fundamental distinction, however, can be made according to their ultimate purpose. One kind is "farm-centred", or in the case of agribusiness, enterprise-centred, although any surpluses might be used to satisfy the non-production needs of workers or farmers (see fig. 3). This is not the purpose of the system, however; it is only a spin-off.

Another system is the "energy farm" unit designed for the production of energy, usually for distribution via conventional means to distant urban markets. This type of system could be expanded into a kind of "public utility" system in order to include a social purpose other than food production, for example, waste water treatment in a manner that simultaneously produces food and reduces the environmental load. Moulik (1985) describes an urban latrine system in India that, coupled with a biogas generator, produces both hot water and street lighting while reducing the sewage treatment problem.

A third type of IFES is the "community focused" system. It seeks to energize daily life in a variety of ways that answer domestic and community needs, such as cooking and sanitation, as well as individual and community productive

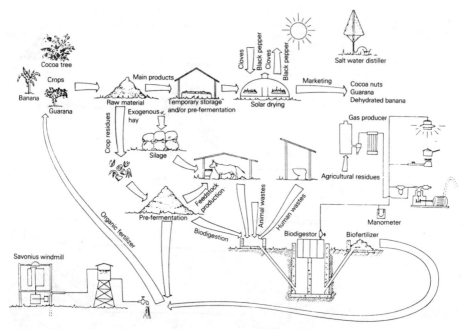

Fig. 3. Flowchart of the integrated production of food, raw material, and energy proposed for the ''agro-energy community'' of Tabuleiros de Valença, Bahia (reproduced from the *Food-energy Nexus Newsletter* 1, no. 1 [September 1984])

needs in agriculture and industry. Moulik (1985) described the features of this type of system as including:
- using a technology mix designed as a minimum cost alternative;
- meeting energy needs not only for agriculture, but also for other social needs;
- maximizing utilization of available bio-resources with minimum harm to the environment;
- benefiting all classes of the community;
- increasing food productivity;
- generating additional employment and off-farm income; and
- requiring minimal maintenance so that villagers themselves can operate and manage it.

It should be noted that IFES do not necessarily offer a fundamental departure from conventional models of ''trickle down'' and ''modernization'' development. Thus those designers who wish to put bio-energy systems at the service of ''another development'' should pay close attention to the local patterns of food and energy provisioning in addition to other basic needs such as clean water.

The same situation applies to the ability of IFES to balance economic, social, and ecological evaluations. Although an analysis shows that environmental degradation can be prevented by using IFES, it remains to be seen whether or not this is actually happening. This raises the question of whether or not IFES that are environmentally sustainable can be promoted instead of energy-intensive agricultural modernization projects that aggravate social and economic conditions.

Analytical Trade-offs

To date, most IFES research that has been done suffers from an engineering bias. Although this provides systems with a solid foundation, such research needs to be complemented with increased contributions from interdisciplinary teams, notably from the biological and social sciences. While most systems take advantage of biofertilizers, there seems to be a "black box" approach to the material that goes into and comes out of biodigestors. There are few studies of the biological, chemical, and physical properties of digested material or of their impact on the soil treated. Little work appears to have been done on the microbiology of biofertilizer. This is particularly true for small-scale systems, notably composts, that have been all but ignored by major research and development organizations (Silk 1988).

There is also a need for nationally or internationally standardized ways of measuring the parameters of biofertilizer systems and of evaluating their performance. Basic comparability is not currently possible because of the differences in the manner of reporting data on IFES. Another problem is that costs can rarely be taken as absolute values but must be related to local, regional, national, and international markets that are constantly changing.

The Chinese Energy Village described by Zhang et al. (1986) is a community-oriented system that seems to transcend some of these problems. Because it was designed in co-operation with the residents of a "natural village", the results allow for a complex, transitional pattern of energy use. Coal, diesel, and kerosene continue to share a place with the new and old renewable energy sources.

Production techniques for both market and subsistence needs are addressed simultaneously by this system. The latter include both family requirements (lighting, cooking, space heating) and social purposes (new fuels). Both productive and non-productive uses are thus served and there seem to be advantages for everyone in the village. Unfortunately, such a balanced system is hard to obtain in real life as IFES are not sealed off or protected from the rest of the economy (and this is increasingly the case even in China).

Most research on IFES has been based on the energy profile because of the much greater experience with energy studies. Indeed, among the early initiatives of FEN was an analysis of the entire chain of food production activities "from the farm gate to the food plate" (Parikh 1985). It naturally paid special attention to the intersections with energy flows, the by-products produced and the losses at each stage, thereby focusing attention on ways to increase efficiency by reducing wastes.

In this context, Alburquerque (1986) identified five basic steps:

1. description of the system to be studied;
2. an energy analysis of the system;
3. identification of possible technological paths for the supply of energy to it;
4. an assessment of alternative technological paths; and
5. a plan for the introduction of the chosen project.

Given the complexity of such an analysis, there is usually a trade-off between the quantity of detailed information one would like to have and the amount one can afford, given constraints on money, personnel, and time. One must thus determine the minimum amount of information necessary to understand an existing agro-energy system well enough to make an intelligent intervention. This leads to the question of how long reasonable baseline research must take at the micro-regional or community level.

The answer, of course, depends on the amount and nature of background data already available. Where baseline data exist, they can certainly be useful, but there are caveats. First, relevant data are sometimes to be found in places where many energy planners would not think of looking, such as nutrition surveys or anthropological studies. Second, all relevant data are not necessarily quantitative. An understanding of the historical background of a community is vital as it is the historical experience of underdevelopment that conditions what the local population perceives as a resource. Deeply-rooted social relations often define access to these resources, which are perceived by everybody but only accessible to some. There is also the problem of studies that have omitted the poorest groups because of various spatial, social, and seasonal biases (Chambers 1983).

With regard to the minimum amount of information needed, it must be noted that local factors are too complex for sweeping generalizations. The practical consequence for the policymaker is that only a very much strengthened, decentralized, retrained, and highly-motivated extension service can possibly monitor the situation sufficiently closely and continuously to provide the necessary data for policy decision-making.

Among the social impacts of IFES, attention should be given to highly seasonal employment patterns, particularly in large-scale sugar cane production, and the poor quality of life of rural families living on vast plantations. The inclusion of other crops and economic activities in these systems would help to provide year-round employment.

A related concern is the consequence of "creating" a wide variety of "new resources'' in rural areas. What will happen, for example, when marginal pasture resources (which may be vital to local users) become potentially valuable as wood fuel plantations or energy farms? More emphasis also needs to be given to the impacts of IFES on women and children.

Regenerative Agriculture

Attention should be given here to the potential of IFES to contribute not only to the sustainability of rural resource-use patterns, but to their ''regeneration''.
    As Freudenberger (1988) has aptly observed:

The word 'sustainable', which is frequently used in reference to new agricultural futures, too often is interpreted to mean that, given necessary resources, even a poor system can be sustained for a long time, provided only that a community has the ability to *obtain* the needed resources. To move beyond this ambiguity, the word *regenerative* is used.
    The idea of regenerativeness goes beyond conceptualisations of conservation, for this latter word usually just conjures up the idea of being careful about using a resource in order to extend its time horizon as much as possible. Regeneration, in contrast, and particularly in the case of agriculture, refers not only to the replacement of the essential resource, but, hopefully, to its enhancement.

In this context, carefully developed IFES could actually enrich the natural resource base by restoring stability and integrity in both the socio-economic and biophysical sense. An analysis of the traditional management system of the Spanish dehesa (Perez 1986) shows the possibility of applying this approach to other regions suffering from desertification and/or conflicts between food and energy production.
    The dehesa is the result of centuries of interaction between farmers and their environment which has resulted in open, savanna-like woodlands with a rich diversity of flora and fauna, thus giving it stability and resilience. It allows sustainable harvesting of fuelwood from periodical pruning and clearing, which also improves the quality of the trees and pastures, thus increasing food production.
    Apart from cereal crops for human consumption, dehesas mostly produce fodder crops to supplement the natural grazing resources. The carefully cultivated oak trees provide not only fuelwood but forage and acorns for livestock fattening in addition to cork, which is used as an industrial feedstock.
    According to Perez (1986), this analysis shows that ''an agro-silvo-pastoral combination producing food, energy and forest products is possible'' and can provide a ''balanced diet and all the products rural populations may need''. This resource-use pattern shows that sustainable resource exploitation is not only compatible with ecosystem regeneration but actually part and parcel of it.
    In hoping that the lessons to be learned from the Spanish dehesa will contri-

bute to improving living conditions in many rural areas of third world countries, Perez (1986) reminds us that "the different environmental and socio-economic characteristics of each region necessitate specific solutions and it must be borne in mind that the dehesa example cannot be transported *per se* to other areas without taking into account these considerations."

Agricultural Refineries[2]

The situation of third world countries can perhaps best be understood by an examination of the experience, both positive and negative, accumulated in this respect by Asian and Latin American countries. The two largest, land-scarce countries, China and India, are still predominantly rural, in contrast with Brazil, where the abundance of land goes hand in hand with premature and very costly urbanization. Both China and India have an excess labour force in agriculture; their problem, therefore, is to promote the agricultural exodus.

The official Chinese policy is aimed at releasing some 150 million workers from agriculture by the end of the century. The slogan is thus: "The Chinese peasant leaves agriculture but stays in the village". As for India, Chakravarty (1987) makes it clear that in spite of the Green Revolution and surplus food production,

agriculture cannot absorb fully the surplus labour even if the existing regional unbalances in agricultural growth are overcome. On the other hand, neither can industry absorb the migrant labour force even with a high rate of industrialization. It would, therefore, be necessary to adopt policies which would generate adequate off-farm employment in the rural areas and small towns so that migration to urban and metropolitan cities and the consequent accentuation of urban social problems could be avoided.

The basic problem, therefore, for all developing countries with a rapidly growing population and increasing rural-urban migration, both permanent and seasonal, is how to promote off-farm migration through "industrialization without depeasantization" (Abdalla 1979). For China and India, this is a necessity, and for a country like Brazil, it is an opportunity to use land productively while avoiding the unnecessary costs of excessive urbanization.

Rural industrialization is usually associated with intermediate or even simpler technologies. But it need not and should not necessarily be so. Recent advances in communication, agricultural, and chemical technologies have made the problem of economies of concentration and scale somewhat obsolete because they allow many secondary and tertiary activities to be deployed in rural areas.

To the extent to which underdevelopment is a "co-existence of asynchronisms", selective modernization, based on an endogenous project, calls for skillful management of technological pluralism. No developing country can sustain across the board the deadly rhythm of accelerated obsolescence imposed by competition on world markets. Neither can it afford technological stagnation. Hence the need to plan sectoral modernization and to design public policies that can offset the homogenizing effect of market forces.

It is in this context that the relevance for rural areas of decentralized IFES based on biomass should be understood. Such industrialization should start with using existing resources and raw materials. Agricultural residues and biomass production are the obvious choices. In tropical countries, this is particularly true for "agricultural refineries" designed to produce a broad range of industrial products from biomass (Munck and Rexen 1985). The key element in this alternative rural development strategy is a process based on valorizing biomass as a dynamic anchor activity around which organizations and other development activities can work.

The "agricultural refinery" concept can be illustrated by the case of sugar cane production. Its central anchor activity is a modern processing industry around which all other activities could be oriented. An average sugar factory with a 2,500 tonne-per-day capacity in India commands an area of about 10,000 to 12,000 ha spread over 30 to 40 villages, employing about 700 to 800 people on a permanent basis and about 2,500 people on a seasonal basis for about 150 to 180 days per year. It also generates secondary and tertiary employment in the service sector. The full potential of such an industry, however, can only be achieved if all of the possibilities of by-product processing are exploited.

This alternative is particularly relevant to Brazil where its National Alcohol Programme is the largest such endeavour in the world, with over 11 billion litres produced per year. But the current low oil prices are causing intense controversy over its cost-effectiveness. The rationalization in the use of existing distilling capacity by improving the productivity of the entire plantation is thus imperative. Turning alcohol distilleries into "agricultural refineries" for overall sugar cane processing is one of the best ways towards this end. This is being shown by experiments now underway in some installations where sugar cane bagasse is being used as an input for cattle raising and electricity generation.

It should be noted that a similar "agricultural refinery" system can be designed for almost all agricultural commodities, such as rice, wheat, cotton, maize, etc. Such IFES would obviously depend on the agricultural and other natural resources available from local sources in the area concerned.

Given recent advances in science and technology, particularly in chemical processing and biotechnology, the exploitation of biomass for IFES opens up new vistas for alternative development. In fact, with the emergence of such a rural industrialization process, and new rural-urban configurations, one can visualize the end of the costly urbanization process and a chance for third world countries to leap-frog into the 21st century.

Food-Energy Nexus and Ecosystem

The 1986 FEN conference in New Delhi (Moulik 1988) shared the interest in methodological themes introduced in Brasilia but placed more emphasis on issues of public participation in the design, implementation, and evaluation of IFES. It also identified a number of policy questions:

1. Have rural energy programmes supported or detracted from equity-oriented development programmes such as land reform and the provision of rural infrastructure and services?
2. Is there a built-in bias in favour of urban systems at the expense of rural systems?
3. Are agricultural extension services giving sufficient attention to resource-conserving methods capable of increasing yields without significant extra energy inputs?
4. Can existing "high output agricultural systems" be modified to include the production of rural energy as well?
5. Is enough being done to promote agroforestry?
6. Can forestry, agriculture, water supply, and energy not be better linked by using a watershed management concept?
7. Is experimentation with bio-energy and other technology for IFES being carried out in the context of the whole socio-environmental systems within which potential users really live and work?

Compared to Brasilia (FINEP/UNESCO 1986), the conference in New Delhi advanced to the point of discussing technical obstacles to retrofitting specific agro-industries; rationalizing specific problematic features of current intensive agriculture in Asia (notably energy for irrigation, soil fertility, pest control) or of bottlenecks in specific national programmes (slurry disposal problems or energy crop choice for reforestation). This suggests that IFES research has reached the stage of "early maturity" that warrants focus on obstacles that stand in the way of quantum leaps in the mass application of such systems.

The emergence of "sideline", "real" or "parallel" markets and activities demands a reassessment of how one carries out "village ecosystem" studies. Although impressive competence at estimating energy flows in villages has been developed over the past few years, there still is little formal understanding of these micro-macro connections. Energy-based definitions of "efficiency" are certainly important, but they need to be complemented with the viewpoints of local people based on their daily lives. Otherwise, there is a distorting influence that leads one away from questions of food quality, welfare, and health benefits of alternative uses of a carbohydrate source such as sugar cane (Seshadri 1986).

Another methodological theme to emerge was the value of "participatory action research" that exploits local knowledge of micro-environments and emphasizes the importance of building on the skills and knowledge that exist in villages.

There is no doubt that everything is in place for large-scale, rapid progress in meeting human needs through IFES innovations. The scale of the potential benefits is enormous: an ESCAP (1985) study concluded that a biomass production/industrialization scenario for India would generate 2,000 million work days of employment and 90 million tonnes of charcoal, serve as a reliable energy base for rural industrialization, and put 20 million ha of wasteland under forest cover, while still being a strictly bankable proposition.

Other elements for rapid progress are also available. As noted above, the engineering side of IFES has clearly reached a degree of maturity with designs, prototypes, and functioning systems now available. Two missing links, needed to make effective use of this experience, have begun to be forged: rapid advances in biotechnology and breakthroughs in the social organization of effective community participation. What is needed is more focus by IFES proponents on integrating their work with these two links.

What is also needed is to translate the political will at a governmental level into the first steps in actually reworking some of the distorting macro-economic and social policies inherited from the past. Such action need not be all that revolutionary but could include some steps towards reforming price, taxation, and subsidy policies needed to implement IFES systems. Another important move would be to give concrete meaning to the verbal commitment to women's equality through such things as day care facilities for rural women engaging in new employment schemes.

A major recommendation emanating from the New Delhi conference was for the establishment of a permanent working group for the study of the Food-Energy Nexus and Third World Development Strategies to be entrusted with the following tasks:

- to organize in co-operation with specialized networks the exchange of current information on on-going research and development projects;
- to promote the preparation of state-of-the-art reports on country programmes and experiences, the potential of new technologies, public policy issues, etc.;
- to organize periodic international workshops aimed at reviewing and evaluating research in progress and also identifying relevant topics;
- to implement comparative research studies;
- to compile a directory of institutions and projects working on IFES and alternative urban development strategies; and
- to seek, on the basis of a pluriannual programme of studies, the support of international organizations and research bodies for the proposed activities.

Conclusion

In order to understand the problematique discussed above and to promote this development strategy, it is necessary to examine the following key planning parameters on a cross-cultural, comparative basis:
— existing patterns of biomass use;
— relevant public policies in relation to industrialization, science and technology, and pricing;
— the techno-economic feasibility of proposed "agricultural refineries" in terms of modern, technological imports, and scale of operations; and
— the organizational and management systems that are required.

Detailed need assessments and inventories of local resources have followed national energy balance studies in a number of countries. There is preliminary experience in the design of agro-energy systems for and with communities. A consensus is also emerging about the types of information needed for such design work, although there are differences concerning "how" to go about getting it. It is also generally agreed that nationally dominant ideas about the nature of development as well as specific government policies (legislation, credit, etc.) have a strong influence over the nature and pace of food-energy development in rural areas.

A number of interesting small-scale technologies are under development in various experimental programmes in several countries. Some of these systems or sub-systems are already being popularized, while others are within a few years of such implementation.

Most experience so far concerns biodigestors. There are still questions of optimal scale for a given purpose, continuity of production, the mix of raw materials, and construction with low-cost materials and local skills. Work on micro-distilleries is well advanced, as is that on small-scale electricity generation using biomass fuels of various kinds. Numerous systems now incorporate gasifiers, but further work is needed on the range of energy values represented by different kinds of vegetable residues.

Small hydro-power and animal traction power sources seem yet to be fully utilized as components of IFES. Aquatic sub-systems, however, have been widely included in experiments and designs, although more data on their costs and benefits are needed.

On the whole, small-scale agro-processing technology seems to be keeping pace with the development of new energy sources and possibilities of conversion. However, there is a divergence between food industries oriented towards export and luxury urban markets and those whose products fit directly into the local diet.

In general, there seems to be little systematic analysis of the possible linkages between new food-energy systems and the institutions representing other basic need sectors, such as schools and health centres. Domestic water supply is the one exception. Intriguing work in a few countries suggests, however, that IFES can be of use not only in meeting the function-specific needs of health centres and schools, but in enabling these centres to become energy centres in their own right through the diffusion of food-energy innovations.

Such possibilities require that food-energy designers situate their work more fully in a "rural development" perspective. This, in turn, suggests the need for more direct involvement by researchers in broader debates about the social priorities implied by national policies. In this way, potential beneficiary groups, such as the urban poor and the rural landless, who currently feature in such systems only as consumers and labour power, might be drawn more directly into IFES. The full potential for techno-social integration, implicit in the food-energy approach, could thus be realized. This would complement the realization of bio-energetic integration and socio-economic integration.

In an analysis of the social impacts of food and energy technologies, Peemans (1987) offered several recommendations for future work in this field:
1. developing a maximum number of synergies between food crops, livestock, fish production, and new sources of renewable energy (biodigestion of wastes, biomass production and treatment) in order to increase the local energy potential for household purposes, irrigation, biofertilization, and transport;
2. focusing greater attention on the role of small-scale rural industries oriented towards the processing of local staples for "export" to urban consumers. This could include the enrichment of local flours with vegetable protein and the production of soybean milk, both of which would help to reduce imports;
3. giving attention to a variety of technologies assessed for their utility in every local context;
4. assessing carefully the institutional framework, including the feasibility of a network of rural and urban co-operatives to support community-based integrated development schemes;
5. promoting the participation of local authorities and central agencies in such a network, in which they could take initiatives and give support.

He concluded that "such a strategy is a slow process and is not a short-term solution to the crisis of the dominant food and energy system. But to help its emergence would probably have more rewarding results in the long term than previous 'hit and run' activities".

4

Alternative Urban Development Strategies

The Urbanization Race

The second major focus of FEN was on ways to increase access to food and energy by the urban poor through encouraging innovative, resource-conserving, and employment-generating urban development strategies. FEN's urban projects shared many points in common with the "resource-conserving" cities of Meier (1974) and the Managing Energy and Resource Efficient Cities (MEREC) programme of USAID (Bendavid-Val 1987).

Urbanization is by far the most important social transformation of our times. In 1800 no more than 3 per cent of the world's population lived in cities, but by the year 2000 urban dwellers will outnumber the rural population. Furthermore, most of the increase will occur in third world cities. Bairoch (1983) estimates that from 1950 to 2025 the urban population in third world countries will have multiplied almost 16 times, from less than 200 million to 3,150 million people (table 1). By comparison, the urban population in industrialized countries multiplied only about five times from 1840 to 1914, the period of its most intensive growth.

The consequences of these trends are assessed in diametrically opposed fashion by the supporters and the foes of large cities. The former emphasize the civilizing role of cities, the high productivity achieved by industries and modern services thanks to their unprecedented degree of concentration, the amenities of urban areas (in sharp contrast to the apparent drudgery of rural areas and smaller centres), and the multiple opportunities for work and self-realization offered to their inhabitants.

The latter insist on the parasitic character of the city, diverting and draining for its own advantage the economic surplus produced by the countryside. They point to the deep disruption of the urban environment with the attendant health hazards, the often appalling housing and working conditions of the urban poor, the endemic unemployment and underemployment, and the social anonymity resulting from sub-human living conditions.

Jacobs (1984) may have gone too far in her unilateral celebration of cities

Table 1. Urban population by major regions in percentage, 1960–2025

Region	1960	1970	1980	2000	2025
World	33.6	36.9	39.9	48.2	62.4
Less developed	21.4	25.2	29.4	40.4	57.8
More developed	60.3	66.4	70.6	77.8	85.4
Africa	18.4	22.9	28.7	42.2	58.3
Latin America	49.3	57.4	65.4	76.9	84.4
North America	69.9	73.8	73.8	78.0	85.7
East Asia	23.1	26.3	28.0	34.2	51.2
South Asia	18.3	21.2	25.4	36.8	55.3
Europe	60.5	66.1	71.1	78.9	85.9
Oceania	66.3	70.8	71.6	73.0	78.3
USSR	48.8	56.7	63.2	74.3	83.4

Source: UN Population Division

as prime movers of economic development. Braudel's interpretation (1979) is much more subtle. Whatever the wonders achieved by cities with the economic surplus that they have been able to concentrate, one should not forget that primitive accumulation was largely through extracting this surplus from the peasantry of today's industrialized countries and their colonies.

At stake for third world countries is the opportunity to transform their condition of "lateness" into an advantage. Modern science and technology associated with a critical analysis of the impasses of industrialized societies should allow third world countries to find alternative patterns of urbanization and to implement them at far less social, economic, and ecological cost.

This is not to say that substantial progress could not be achieved in third world countries through more traditional, Western-style methods. But the magnitude of the financial effort required makes this proposition unrealistic in the present political context, even if the volume of necessary investment does not exceed the theoretical possibilities of the world economy.

Yet it may also be said that urbanization cannot be carried out along the old lines, following the model of the large cities of the industrialized countries. To put it more precisely, in third world countries the pattern of city growth – reflecting that of the industrialized countries – and the resultant increase in the speed of urbanization, render it practically impossible to cater to the basic requirements of the majority of third world residents.

To simply house the additions to the urban population between now and the end of the century, the equivalent of over 600 cities of one million residents each would need to be built. An exhaustive study of the metabolism of Hong Kong by UNESCO's Man and the Biosphere programme (MAB) concluded that the energy costs of constructing and operating such cities between 1978 and the year 2000 would be five times higher than total world consumption of energy in 1973 (Newcombe et al. 1978). Future cities must therefore be resource conserving. The pioneering work of the Washington-based Institute for Local Self-reliance shows that the scope for energy conservation in the urban setting is, in fact, quite high (Morris 1982).

Finally, one ought to mention the lack of appropriate policies to implement such an ambitious construction programme and to apply resource conservation principles in an equitable manner. Hardoy (1982) rightly pointed out that in urban studies a certain degree of consensus exists about macro-problems, but nothing of the kind happens with respect to micro-problems which affect low income groups more directly. There is a visible gap between the effort made to arrive at the diagnosis and the neglect in creating the necessary policy instruments.

The Real Economy of Cities

Special reference should be made here to the FEN project with the Colegio de Mexico, "Going beyond the formal/informal dichotomy" and proposing a more comprehensive analysis of the "real" urban economy (Sachs 1987). This was done in the context of seven large cities in both the South and the North, exposing the web of interconnected markets, non-market activities, and the multiple forms of state intervention. An overview of this project can be found in Sanchez (1988).

The ways in which market and non-market economic activities combine are quite complex. They constitute the fabric of the "real" economy. Non-market economic activities should not be viewed as a residual category designed to disappear with technical progress. On the contrary, their share in terms of time allocation may increase, *pari passu*, with the reduction of the working time against wages, as a matter of deliberate choice. The market/non-market dichotomy therefore offers a useful starting point to move in the direction of a development theory not exclusively based on the categories of the market economy and on monetary metrics.

The informal sector is often referred to as "hidden" or "invisible". A better term would be "statistically unrecorded" as most of the activities in the market segments encompassed by these names are conducted out in the open. Neither the "formal/informal" nor the "open/hidden" dichotomy offers a suitable framework to describe the latticework of the real economy. Moreover, they lend themselves to statistical manipulation, as both the informal and hidden sectors are in reality residual categories.

By definition, they include everything that has not been specifically recorded as belonging to the narrowly defined organized sector of the commoditized economy. By postulating that all those who do not find a regular job in the organized sector are absorbed by the informal economy, it becomes possible to assume away the problems of unemployment and underemployment, as well as to play down the disruptive social consequences of the emergence in industrialized countries of a "two-speed" economy: a highly performing and competitive sector open to a minority, and a residual one for the rest of the population.

The large cities thus prove to be cities for the elite which, at best, function as mechanisms for the regressive redistribution of investments and wealth. The poor end up by paying more dearly than the rich for the basic services, without which life would be impossible. The extreme case is illustrated by cities like Karachi where the inhabitants of poor neighbourhoods have been known to pay itinerant water-bearers 20 times more for their (questionable) water supply than is paid by the inhabitants of the rich neighbourhoods where running water is supplied (Ward 1979).

In contrast to the commoditized economy, household activities are situated both outside the labour and product markets, although household consumption obviously consists of a bundle of goods and services that are both purchased on the market and self-produced.

In addition, the non-market segment of the economy also comprises the social sector, consisting of all collective activities organized outside the market by neighbourhood and community groups, citizen associations, and, in some cases, co-operatives. Their common trait is that they are founded, just like the household sector, on the principle of reciprocity: the donation of an unpaid productive activity, deemed of social interest and matched by free competition of goods and services collectively produced.

A closer look at the real economy of the city (including the multiplicity of interconnected markets, ranging from the legal to the criminal, non-market household activities, and state intervention through to subsidies, rationing, distribution of goods, and provision of services) shows that many of these resources, not accounted for in official statistics, are being intensively used by people to build their homes, produce some food for self-consumption or sale, and transform recycled waste into saleable commodities.

A Bootstrap Operation

What will happen now, in times of crisis? For the bureaucrats from the ministries of planning or finance, the answer is clear: austerity budgets compounded by the requirements of foreign debt servicing mean that the so-called "non-productive" investments and collective consumption must be cut to the bone.

Whatever the outcome of the debate between the partisans and critics of International Monetary Fund-style policies, one thing is clear. Difficult as it is, the situation is not entirely stalemated so long as there are idle, underutilized, or dilapidated physical and human resources that can be used to produce socially desirable goods and services without violating the prevailing budgetary restrictions. A bootstrap operation would be based on reducing waste in order to increase the resources available for development. It would be aimed at improving living conditions in poor urban areas through grassroots community action without waiting for massive funding from outside. Such an operation has, of course, obvious limitations. It cannot by itself solve the economic crisis and generate enough jobs to reabsorb the backlog of unemployed. Furthermore, under no circumstances should it be used as an excuse for the authorities at local, regional, and national levels to shirk their responsibilities in this field.

On the contrary, grassroots action must be actively supported. Self-reliance does not necessarily mean self-sufficiency any more than an inward-looking development strategy leads to delinking (Sachs 1984). The complexity of the modern world cannot be tackled by decomposing it into an archipelago of self-sufficient communities, be they rural, urban, or rurban. But self-reliance in moral, political, and intellectual terms makes people resourceful and confident: they assume their situation instead of taking a passive approach; looking around them, they end up by identifying resources in their own backyard that can be exploited to bring some relief to their plight.

The FEN message is thus not one of unqualified optimism and romantic idealization of grassroots movements and vernacular technology. Its purpose is to focus on a potential operating margin in a situation otherwise completely deadlocked.

Resource-conserving Cities

Economists tend to look at cities as the site of many enterprises, whose concentration creates both positive and negative externalities but requires a costly infrastructure. Human ecologists have been advocating, without much success to date, the study of cities considered as ecological systems (Boyden 1984). However, most of the studies conducted within the MAB programme on urban systems deal with the impact of cities on the natural environment and its food-producing systems, or describe in detail the energy flows inside the city.

The approach followed by FEN was somewhat different. For analytical purposes, it considered the city as a predominantly artificially created ecosystem with paradigmatic analogies in relation to natural ecosystems. Such a perspective emphasizes the actual and potential interrelations and complementaries between different human activities conducted in the cities (see fig. 4).

Fortunately, in most cities the backlog of untapped opportunities for transforming waste into wealth is very large indeed. A city should thus be regarded as an ecosystem with its own potential of latent, underutilized, misused, or wasted resources.

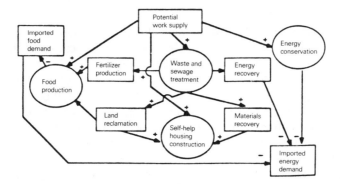

Fig. 4. Impact diagram of the food-energy nexus in the city (reproduced from Sachs 1986)

Whenever possible, loops must be closed and residues from one production transformed into the inputs for another. The urban ecosystem thus appears as a vast potential of physical and human resources to be identified and used to improve the quality of urban life, especially that of the poor.

The analogy between the urban and the natural ecosystem as resource potentials is valid to the extent to which resources do not exist as such. They are but portions of the environment, natural and urban, that people learn to use for a specific purpose. Knowledge about the environment, or if one prefers, culture, is thus an essential component of the very concept of resources. And resourcefulness, the ingenuity of transforming into resources things around oneself, is an important cultural asset for a more self-reliant development. The issue is not to give up access to any other resource, but to make the best possible use of local opportunities, combining them with the flow of external resources to the extent to which they are forthcoming.

In this respect, waste treatment could play a role by providing additional productive employment and generating inputs into food and energy production as well as self-help construction. Although the margin for energy conservation in third world cities is less than in the more energy-intensive cities of the North, much can still be done with respect to energy consumption in industries and cities. Self-help housing also offers significant potential.

Genuine Participation

Urban ecodevelopment calls for a detailed and concrete knowledge not only of the city's latent resources but of its pressing social needs. Identifying and matching them requires the constant and effective participation of grassroots organizations and citizen movements, because of their daily involvement with the territorial specificity of each neighbourhood. Planning for urban development certainly requires going down to this scale, even though it cannot be performed exclusively at the lowest level of disaggregation.

The meaning of participation must be spelled out because of the frequent abuses of this term. The formal and passive association of community organizations with policies initiated by the authorities constitutes, at best, a pale imitation of what is required and, at worst, a cover for authoritarian regimes. Genuine participation ought to be measured by the power of initiative gained by the community, the room for real-size local experiments, and the degree of symmetry in the relation between the citizens and the different levels of government.

Existing mechanisms for concentration and conflict resolution, the nature of the planning process, and access to the media should also be considered in addition to the capacity of community organizations to find a balance between their roles as critics of the existing order and as proponents of constructive solutions. There is, of course, room for both.

In other words, it is necessary to look at the place for genuine participation provided by the formal interplay of institutions and at the actual unfolding of the political process, both in its party and non-party manifestations. The interaction among the actors of the development process is closely related to the articulation of development spaces – local, regional, national, and international.

Fear is often expressed that insistence on local development may produce a perverse effect in the form of exacerbated parochialism. Such a danger certainly exists, but to refer once again to Morris (1983), the inward orientation of local self-reliance may be compensated for by the outward orientation of modern communication systems. Communication can play a dual role as a positive means of social control over the working of the political and administrative systems and as a tool for horizontal networking of communities interested in exchanging experiences, technologies, and products of culture.

As Finquelievich (1986) observed in the case of Latin America, governments have attempted for several decades to deal with access by the urban poor to food but until recently their measures were strictly institutional and excluded public participation. They consisted primarily of food subsidies, price controls, and food distribution through state-controlled channels. Khouri-Dagher (1987) noted that in Cairo (and presumably elsewhere), such programmes could result in considerable waste of resources and could wind up subsidizing the rich more than the poor. For El-Issawy (1985), such shortcomings can be cured by modifying the system.

Since the 1970s, self-production of food, joint purchases, community gardens, and other initiatives have begun to appear, gaining ground and government support in response to worsening economic conditions. These actions helped to improve the living standards of the urban poor by relocalizing food production and strengthening local control, confirming that grassroots organizations can have a far-reaching impact. This is due, in part, to the fact that such groups have "ingenious" projects and also succeed in getting governmental co-operation, especially at the local level (Finquelievich 1986).

The higher degree of success observed at the local level is attributed to the fact that a municipal network is easier to manage and there is greater interaction between local authorities and community organizations. Such co-operation has the advantage, on the one hand, of receiving government support and relying on the efforts of those people that it serves. Because community groups are more in touch with the real needs of local people, projects involving governments, NGOs, and users have been the most effective.

On the other hand, "alternative" projects based only on community support face a number of problems, not the least of which is their reliance on voluntary labour. Unless the initial motivation is extremely high and creates considerable momentum, such initiatives usually do not last. Such was the case of the community kitchens established in Osasco, Brazil, that were designed to reduce nutritional problems by providing communal cooking facilities.

Reporting on these kitchens, Cardoso (1985) noted that alternation between enthusiasm and indifference may occur with such projects because just a few people are always in charge of the work. While the community kitchens that were associated with an active women's group achieved some continuity, those that relied on a hierarchical form of organization had problems getting people to co-operate because of a lack of cohesion.

Kitchens that were introduced through initiatives taken by the municipal government caused problems in communities where people were already organized at the local level. As Finquelievich (1986) observed, the creation of social organizations in which people assume responsibility for themselves and their community is often as important as the solutions that they reach.

Cardoso (1985) concluded that:

It is thus important to avoid simplifying the reasons for the lack of participation by resorting to a vague notion of "welfarism" which, while criticizing the initiative taken by the authorities, leaves the burden of solving the problems of the underprivileged on their own shoulders. It is undoubtedly true that the presence of the State in inducing a participation process raises new questions, but for this very reason it must be carefully observed. In this new context, the challenge is to promote self-management regardless of the private or public origin of the resources involved.

Communication for Urban Self-reliance[1]

An analysis of the degree of communication within and between institutions and individuals involved in urban self-reliance revealed that most of the many activities undertaken in this field are generally done in isolation. The GRET/FEN project in this field concluded that such communication can be defined at two different levels: "horizontal" communication at the national or international level between similar groups and "systematic" communication between all actors involved in urban self-reliance.

The needs expressed are not only multiple but are evolving as a result of rapidly changing relationships ranging from repression by the powers that be to collaboration between all actors involved. It was found that most needs were in the field of practices rather than purely technical information. But the production of information on existing practices is complicated by the different contexts, social and cultural backgrounds, language barriers, and different degrees of development of such projects.

For many people, the actual means of communication are a secondary issue when compared to more crucial questions such as: What should be communicated? How should information be screened? Who should information be communicated to? For the time being, it was concluded that most means of communication are poorly adapted to existing situations either because of their cost or incompatibility.

The role of networking, although poorly defined, was identified by most individuals and institutions contacted as a potential answer to communication needs for urban self-reliance. Those involved in networking generally share three points in common:
1. a willingness to communicate;
2. common goals; and
3. a preference for decentralized or horizontal communication.

The activities of such networks can be classified as follows:
(a) exchanges (of information, people, tools, etc.);
(b) joint action;
(c) mutual support;
(d) lobbying; and
(e) common publicity.

The major conclusion reached by the GRET/FEN project was that the vitality and dynamism of a network does not rely on a centralized institution which acts as a secretariat but on the catalytic action of individuals who facilitate connections. These people are able to see what might be useful for another colleague or are able to identify quickly common ground for other members of the network.

In combination with the International Foundation for Development Alternatives/FEN directory (Cordova-Novion and Sachs 1987), such activities contributed to the identification of more than 200 institutions and projects involved in urban self-reliance. Many such innovations are already happening. Information about them should continue to be collected, evaluated, and disseminated as widely as possible, not as models to be followed but rather as stimulation for the social imagination of all those who live in the cities of the third world.

The recent creation of RECEM, a communication network on municipal experiences established by the Fundacao Faria Lima in Sao Paulo, is a welcome step in this direction.[2] The next stage should consist of involving international organizations, networks of research institutions working on urban problems, and NGOs active in this field, in a common effort to establish a systematic South-South flow of information. Mutual knowledge among third world countries is a precondition for meaningful collaboration and, ultimately, greater collective self-reliance.

Lopsided Modernization

Complex and diverse, urban environments combine elements of natural and entirely artificial environments. They juxtapose modern factories, lavish residential quarters, and suburban expressways with decrepit sweatshops, sprawling slums, and antiquated public transportation. The same city provides an array of environments for different groups, a multiplicity of ecological niches ranging from cozy to uncomfortable, from healthy to filthy, from safe to dangerous, from friendly to hostile. These are multi-layer towns, often with a marked spatial separation – garden cities for the rich, shanties for the poor – resulting in what amounts to an apartheid society.

It should be remembered that the urban poor are the main victims of environmental disruption. In addition to living in conditions of squalor subject to the pollution of poverty, they are also the most exposed to the pollution generated by the lavish consumption patterns of the urban elite, including the increasing affluence of the middle class.

The bridge for many is provided by the TV soap operas with their consumerist message: watching them from a shantytown is fairly surrealistic. It forces one to think about the latent explosiveness of the situation in the dual cities of the third world produced by lopsided modernization.

At the bottom of the environmental plight of the urban poor majority lies the imitative growth strategy carried through social inequality and fuelled by the Western modernization model with its attendant features: private consumerism rather than development of collective consumption, primacy of individual careers and lifestyles over social concerns, strong preference for the present with little or no preoccupation for long-term effects, privatization of profits, and collectivization of the costs and risks.

Whatever the judgement one may have of this model in the context of industrial countries, its transposition to less developed countries creates additional distortions: scarce public resources are used to the exclusive benefit of the modernized elites, while the urban majority is deprived of access to basic amenities and decent housing. The urban poor have thus the worst of both worlds and cannot expect much from the caricatural imitations of the Welfare State, too poor to offer effective remedial action.

Even during the heydays of rapid economic growth, most large cities of the third world did not succeed in expanding their infrastructure and basic services, *pari passu*, with the increase of their population. They are now left, in a period of crisis, with a huge backlog of unattended basic needs, or to use the eloquent Brazilian term, a huge "social deficit". The prospect for the next decade or so is grim indeed. Cities will continue to grow with the influx of "rural refugees" and all those who consider, rightly so, that in spite of the odds, large cities remain the locus of the last hope. They offer, in the words of the 19th century French historian, Jules Michelet, at best, a life lottery[3] and at worst, a place where they can expect some *panem et circenses*.

If the life-lottery analogy is correct, one could expect that metropolitization will advance at an even greater speed than urbanization. People will tend to go to the capitals and large cities because of the illusion that more winning tickets can be found there. When news of an industrial boom spreads through a country, for every new job offered, several newly-arrived candidates apply in addition to those already living in the town. When recessions sweeps a country, moving to the largest city appears as a solution of last resort. So, whatever the ups and downs of the business cycle, migration continues.

Without attempting to resolve the controversy of whether cities are "parasitic" or, on the contrary, "generative" (both cases exist with all sorts of intermediary positions), a severe resource squeeze is inevitable unless bootstrap operations can be devised.

There is considerable temptation to postpone all actions aimed at reducing environmental disruption under the pretext of lack of resources. Business should be continued as usual, meaning by that, savage growth first and remedial action some time later. This is a questionable attitude on ethical, social, and even economic grounds. It begs the question of diachronic solidarity with future generations, assuming implicitly that the plight of the urban poor is a necessary cost of "progress" and ignoring the evidence that preventative measures are almost always more economical in the long run.

Focusing on greater food and energy self-reliance thus offers a starting point to design and gradually implement urban strategies inspired by the ecodevelopment approach. Its possibilities should not, however, be overestimated. Much more is needed to overcome the present crisis, since the urban situation cannot be disassociated from what is going on in the countryside. Part of the urban crisis is due to the stream of immigrants from rural areas, where they are unable to earn even the most miserable living.

The investment required to accommodate them in the cities is far greater than the outlays that would be necessary to provide them with agricultural and related jobs if access to land and other resources were only made possible in rural areas by appropriate institutional measures. An overall development strategy is clearly a precondition to tackle successfully the urban and rural situation.

5

Urban Agriculture

A Utopian Dream?

Ever since people have been living in cities, small but highly valued quantities of food have been grown in cities. Apart from historical references to ancient civilizations, however, there is relatively little documentation on the actual extent of urban agriculture in either the near past or the present.

This lack of information on urban agriculture is supposedly due to the difficulty of recording its impact and the absence of a concerted effort to publish research findings in this field. But the fact remains that urban agriculture is often simply ignored, if not marginalized, by those "experts", officials, and residents who could be its biggest promoters and beneficiaries. As a result, the potential of urban agriculture as an alternative to increased agricultural production, more food subsidies, and improved distribution and storage systems is not well understood.

What does this mean for the future of urban agriculture? Is there, in fact, a contradiction between cities and agriculture? Is it simply a utopian dream to suggest that urban areas could really help to feed their inhabitants? Should practically all agricultural development assistance continue to be directed towards rural areas?

Without resorting to highly artificial and expensive food production systems, there is clearly little prospect of growing staple foods, such as wheat and rice, in urban areas. There is, however, certainly room for banana and palm oil trees in third world cities, where these crops clearly show the potential of using urban resources for food production. According to Soemarwoto (1981), urban agriculture can provide some residents with up to 40 per cent of their recommended daily allowances of calories and 30 per cent of their protein needs, including most of the vitamins and minerals crucial to their health.

If such a contribution could be directed entirely to the nutritional needs of the poor, urban agriculture would be of vital importance in third world cities. Its economic and recreational attributes also make it of interest to the more privileged residents of cities anywhere in the world. Indeed, until the middle of this century, "most urban areas around the world produced a significant amount of food and other items required by local residents. Production was not limited to the urban fringe but included substantial yields in home gardens within the cities themselves" (Wade 1981).

During the post-war period of economic growth, however, higher incomes and much cheaper food led most people to abandon home food production in many Western countries. Urban sprawl also reduced the availability of good agricultural land on the fringes of cities around the world. In many third world countries, land that did remain available near cities was often dedicated to growing cash crops for export or for the urban élite.

Survival in the City

Although most people in Western cities have lost their links with the land, this is not the case for the people who are now streaming into third world cities. While many of these rural refugees may not like the idea of continuing to work the land, the fact remains that they do have the survival skills to produce food on their own, given access to the necessary resources.

In two poor neighbourhoods of Bombay, Panwalker (1986) found that rural migrants who came to the city to find a job considered farming to be a demeaning activity that they had left behind. This attitudinal barrier is further complicated in many African countries where the traditional role of women in subsistence farming reinforces the reluctance of men to engage in non-commercial food production.

Deelstra (1987) noted, however, that such survival skills could also be used to facilitate the transition to urban life. By producing some of their own food, immigrants can gain more confidence in themselves, leading to increased security and better integration. Urban agriculture also contributes to increasing one's social contacts, a prerequisite for success in the city.

This, in fact, appears to have been the case at least in Zaire, where Streiffeler (1987) observed that men adapted all too quickly to city ways. When it became evident that urban agriculture could be used for intensive, lucrative production of vegetables as cash crops, the attitudinal barriers vanished overnight and the men displaced the women. Although commercial ventures are traditionally a masculine role, these men generally had little experience in growing vegetables and thus required training courses. One may assume that the women had to grow their own crops on poorer or more distant plots.

Another area where urban agriculture can make a significant contribution is during the "hungry season" between major harvests. As the stocks of stored food run out, and before new crops have matured, the small output of permanent gardens can play a major role for this limited time period. Such produce may not be among the most preferred, but there is always something to eat.

Living Food Stalls

It is estimated that between one-third and one-half of all families in the third world have gardens which, after centuries of traditional cultivation practices, can be very complex, well-balanced affairs. In Javanese, they are known as "living food stalls". Often near to or adjacent to the house, these systems are permanent and sustainable and require low-inputs once established.

Gutman (1987) concluded that 100 sq m of intensively cultivated land in Argentina could supply the vegetable needs of a family of five and that this would require about 1 to 1.5 working days per week. A successful garden in Buenos Aires could thus save between 10 to 30 per cent if not 40 per cent of the cost of an appropriate diet for a family, representing savings of 5 to 20 per cent or more of its total income.

In one of the few publications in this field, Brownrigg (1985) made the point that home gardens in third world countries produce not just vegetables but often plants that provide starch, fruits, herbs, flowers, and medicines as well as fuelwood. In many cases, small livestock are also raised with their wastes helping to fertilize the garden whose wastes, in turn, help to feed the livestock. Yeung (1987) noted that although urban forestry for fruit production is rare in Asia, 25 per cent of the "large number" of street trees grown in the Indian city of Bangalore bear fruit.

Although there is little data available on the nutritional value of gardens, many such projects in the third world are to be found under the responsibility of health rather than agriculture agencies. Indonesia and many other Asian countries have used urban agriculture programmes to improve vitamin A levels in diets (Yeung 1987).

The economic value of home gardens has received some attention but most of it has dealt with the straight market value of the produce, which is difficult to assess in those areas where most food is self-produced or exchanged. While most people normally save money by growing their own food, this is particularly important in the third world where the food budget accounts for a much higher proportion of total family expenditures: often 50 per cent and reaching 70 per cent for urban families in India, compared with 25 to 30 per cent for the average American family. Another consideration is that with gardens ensuring part of a family's subsistence, there is a greater chance that people will risk investing in other income-generating activities.

Kleer and Wos (1987) noted that this implies the need for a different method of calculating economic benefits, given the fact that it involves marginal labour. In a detailed study of urban agriculture in Poland, they concluded that such production also enters the "broadly understood" market and enriches it.

Wade (1987) noted that surprisingly high yields have been obtained in small gardening projects developed in many countries. Production ranges from 6 to 15 kg of vegetables per square metre or the equivalent of 66 to 165 t/ha compared to 1977 FAO averages of carrot production of only 22.6 t/ha. Very high yields are also reported for organic gardening and intensive aquaculture projects of research institutes and individuals working in temperate climates. No comparable effort appears to have been made for tropical countries despite the fact that the growing season is longer.

The potential contribution of urban agriculture to the food supply of the urban poor is thus difficult to assess. As an order of magnitude, we may take the estimate quoted by Morris (1983) on the basis of a six-month growing season: a full balanced diet can be derived from 2,500 sq ft per person or a little more than 0.02 ha for a family of five. A 200 sq m garden would in this case certainly help the family to make ends meet by providing one-fifth of its optimum food intake or much more in terms of the present low nutrition standards.

Accordingly, about one quarter of the population of Sao Paulo (about 425,000 families) would require about 8,500 ha. This figure ought to be compared with the 60,000 ha of empty land inside the Sao Paulo municipality. Of course, smaller kitchen gardens of 50 to 100 sq m are also foreseeable.

New Types of Dialogue

Unlike capital- and energy-intensive agricultural development projects, urban agriculture does not depend on big budgets. It must be made quite clear, however, that urban self-reliance does not mean less work for the state or international agencies but a different kind of work. What is needed is the development of policies that support such initiatives rather than institutional barriers that block them. This is a crucial issue that must lead to new types of dialogue between communities and local authorities.

For a variety of reasons, food is rarely grown in cities by corporations or institutions but rather by household members at or near their place of residence or work. Locally organized gardening activities also avoid the problems of marketing and motivation that beset "top-down", official gardening projects. The latter are rarely as successful as community-based projects, especially if care is not taken in hiring enthusiastic, well-trained extension workers from the local area.

In an analysis of a successful urban agriculture project in Zaire, however, Streiffeler (1987) concluded that the enabling factor was perhaps the likelihood that the expatriate organizer was not compromised by the local power struggles and intertribal conflicts, and that he was not suspected of having self-interest in exploiting the poor.

Sanyal (1986) noted that the evolution of government positions on urban agriculture – from outright rejection, through benign neglect, to lukewarm and occasionally enthusiastic encouragement – will face stiff resistance from the many bureaucrats who played major roles in shaping the industrialization strategies of their countries. These "modernization" experts and their supporters generally oppose any efforts perceived as threatening this goal. As Sanyal concluded, "they would rather see the urban poor either sent back to the rural areas" or settled in new cities.

As Kleer and Wos (1987) noted, the majority of practical problems occur at the grassroots level and the attitude and support of local authorities can be decisive. Streiffeler (1987) refers to the "scorn, indifference and official disdain" accorded by local officials in Zaire (and elsewhere in Africa) to "traditional" activities such as urban agriculture.

La Rovere (1986) described the programme of the Rio de Janeiro state electricity utility (LIGHT) that promotes urban agriculture on its right-of-ways. This is done to reduce maintenance costs (notably herbicide or manual weed control) and to increase food supplies for its employees as well as for local residents and schools. But its main purpose is to establish and maintain a good working relationship with the community in order to increase the chances of expanding transmission corridors without resorting to expensive underground lines.

Gardens as a Catalyst

Apart from its contribution to community development and crime reduction programmes (notably in North American cities), urban agriculture has many other indirect benefits. The conversion of vacant lots into productive green space can help to moderate the micro-climate by reducing noise and dust levels in addition to improving oxygen production through photosynthesis.

Green spaces are also excellent hydrological membranes because the well-worked soil of gardens has high infiltration rates in contrast to the compacted soil of other urban green spaces, which also have the disadvantage of high maintenance requirements. Deelstra (1987) noted that healthy hydrological systems are instrumental in reducing surface runoff and flash floods. They also contribute to stabilizing water tables in urban areas which, in turn, helps to ensure solid building foundations. Replacing unsupervised garbage dumps with garden plots also reduces the risk of groundwater pollution.

Urban agriculture can also be a catalyst for recycling programmes. This is already the case in Calcutta, where vegetables thrive in "garbage gardens" based on that city's highly degradable waste (mostly organic matter, ash, and dust) that is painstakingly sorted by thousands of garbage pickers. The marshes around Calcutta have also long played a role in provisioning the city with fresh fish and ducks that are raised on these natural sewage treatment ponds that have so far proven remarkably safe (Furedy and Ghosh 1984).

By organizing the recycling of organic wastes one also facilitates the recycling of other materials, which not only reduces pollution but generates employment and conserves energy. This has already been partly achieved in Lae,

Papua New Guinea, where a gardening programme led to municipal composting, which now produces sufficient compost for all city gardens and parks in addition to a surplus that is sold to commercial farmers (Wade 1987). This has resulted in a 10 per cent reduction in solid waste disposal needs.

A growing problem with the production of compost, however, is the increasing quantity of heavy metals and other non-biodegradable materials contained in the garbage of industrialized countries. It is not too late, however, particularly for third world cities, to implement source separation programmes to ensure that solid wastes can be recycled and organic wastes safely composted.

The production of more food in urban areas, particularly freshly consumed perishables, also reduces the need for environmentally costly and economically expensive transportation of food, let alone energy-intensive processing, packaging, and storing. Gardening projects could also act as demonstration programmes for new species and techniques for commercial farmers, for new food preservation and storage techniques, and for new cooking tools and methods (which are proving important in the post-harvest food cycle).

Gardens or Warehouses?

The major constraint facing urban agriculture is the availability of land. While there are significant amounts of unused land even in major cities, the problem is one of access and then tenure. Even when food production has been started, it does not rank high among the priorities of developers and politicians. The famous Matalahib Community Gardens in the Philippines – which during their peak supplied about 400 poor families with 80 per cent of their vegetables – were replaced with a warehouse.

Little research is available on who urban agriculture would threaten, if indeed it would threaten anyone. Streiffeler (1987) concludes that it is primarily the big merchants (and presumably big farmers) with their own means of transportation who stand to lose the most from urban agriculture as small rural farmers lack access to urban markets. Urban agriculture may thus also have a role to play in encouraging people to form their own production and distribution systems in order to avoid the middleman, such as has happened in Rio de Janeiro (La Rovere 1986).

But should even these farmers see the development of urban agriculture as a threat to their business? If the produce of urban agriculture is consumed by the urban poor who would otherwise go hungry because they cannot afford to buy more food, such farmers should not be affected. The time invested by the poor in urban agriculture would simply mean less time spent on other activities, which could include scavenging, begging, or stealing. Growing food within the city would thus make a material and moral difference for the lowest-income stratum, if not for the city as a whole.

Time does not appear to be a major constraint as much of the labour required for gardening is often provided by the young and the old who have few other opportunities for remunerative work. This may not be the case for the poorest of the poor, however, who being malnourished and poorly housed, simply lack the energy to invest in what appears to them to be a risky, long-term venture.

In most cities, possibilities exist for greatly expanding urban agriculture by applying fiscal measures discriminating against owners of large tracts of land kept idle for real estate speculation. It should be remembered that putting an urban plot under cultivation is, after all, a fairly reversible decision. All the land reserved for the future expansion of the city could meanwhile produce some food. Individual backyards, schools, factories, and public land unsuitable for construction (e.g. communication and transportation right-of-ways) offer a permanent opportunity for gardening.

Other Obstacles

In one of the rare observations on the attitudinal obstacles to overcome, Tricaud (1987) noted that there is a problem with the various specialists involved in urban land management. They tend to see only the single aspect of urban agriculture related to their own field and regard the other various functions as competing interests.

Yet urban agriculture could fulfill most of their requirements if landscape architects, urban planners, economists, sanitary engineers, and agronomists were to co-ordinate their efforts and make a few compromises. The problem, of course, is that urban agriculture suffers from split-incentives and the lack of an integrated agency that can overcome these sectoral and temporal problems.

For Sanyal (1986), the most basic requirement for facilitating urban agriculture is proper cadastration to record the boundaries, ownership, value, and other attributes of land and buildings within the city. This implies not only a willingness on the part of owners, renters, and squatters to subject themselves to registration (with all the benefits and risks that this entails), but also the existence of an infrastructure of surveyors, lawyers, and administrators in addition to an acceptable judicial procedure to arbitrate land-use and ownership conflicts. This would require stronger municipal governments backed up by adequate revenues independent of central authorities.

Wade (1987) emphasized the almost universal importance of government support in providing land and water sources in addition to an initial supply of planting material. Seeds and seedlings are often not readily available in the community, are too expensive, or are simply too old. A problem with supplies donated from outside is that they are often inappropriate for local growing conditions and/or cultural preferences. The most successful strategy has been collecting seeds from discarded market produce.

Effective technical assistance, including soil testing, is also needed for urban agriculture. Unfortunately, many of the NGOs skilled in community organizing have little expertise in horticulture. As Streiffeler (1987) observed, urban agriculture cannot be considered as simply the continuation of the old habits of rural immigrants. There are fundamental differences between the extensive slash and burn agriculture often practised in rural areas and the more intensive requirements of fixed farming in the city. Another drawback with urban agriculture is that the "survival margins" of urban farmers are so small that they cannot afford to experiment with new crops or techniques (Richards 1985).

Apart from the poorly understood problems of fertilization for urban agriculture and the use of grey water for irrigation, Streiffeler (1987) also noted that work is needed on problems of protection against the tropical sun, torrential rains, and ravenous pests that complicate agriculture in the third world. There is also a lack of tools, particularly those needed to work soil hardened by drought and erosion.

Wade (1987) suggested numerous initiatives that could be used to promote urban agriculture and to improve access by the urban poor to other food supplies. These include granting tenure to squatters to encourage long-term investments in urban agriculture; requiring all new housing construction to provide adequate facilities for home gardening (including balconies and rooftops); and providing tax benefits to individuals or companies who "lease" their land for community food production.

What is needed is a re-orientation of existing resources and programmes to include agricultural and food planning in urban areas. This implies more flexible land-use, water, and waste management. Production alone, however, cannot solve all of the problems: new efforts are also needed to improve food distribution, preservation, storage, and processing of food. Nutritional habits should also be improved by promoting the consumption of traditional fruits and vegetables. Efforts are needed to "debunk" the myth that imported, Western produce is better than locally produced food (Wade 1987).

Conclusion

Part of the problem is that urban agriculture, at least in Africa, has simply "been developed to the point of enabling people to survive" (Streiffeler 1987). While critical of local authorities in Africa, he concluded that Western countries must also share the blame for their unwillingness to fund urban agriculture programmes.

It is clear that urban agriculture cannot replace other strategies (thus income redistribution programmes must continue), but it has the advantage of generating independence. It uses many of the principles of self-reliant, local development based on initiatives that can be undertaken directly by local people using resources already available in the community. Urban agriculture also establishes direct links between the actions and outcomes while minimizing the risk of benefits being diverted to more powerful urban groups.

Gutman (1987) also noted that "urban agriculture helps to create and re-create the links of communication and action, of self-organization and initiative, that are absent in the traditional systems of income redistribution".

Nevertheless, one must plan for high failure rates, with slow growth levels and frequent abandonments. Gutman (1987) estimated that it would take up to ten years for urban agriculture to reach 20 per cent of the households in greater Buenos Aires. This compares with the six months that it would take for a food distribution programme to reach the same number. It must be acknowledged, therefore, that urban agriculture is not a short-term solution and that it may not necessarily benefit the lowest income groups.

6

New Rural-Urban Configurations

Decentralized Industrialization

Hardoy and Satterthwaite (1986) have questioned whether recent trends in urbanization are a useful pointer to the urban future or a sign that the urban age, on the contrary, is coming to an end. Is it possible to conceive of development without urbanization?

A common model for urban change is as unlikely as a common model for economic change. This underlines the need to assess different urbanization patterns using such concepts as core/periphery relations, generative and parasitic urbanism, opposition between rural and urban lifestyles, and the urban bias.

Policy-oriented studies are also needed to develop industrialization and urbanization patterns best suited to the specific conditions of each country or region. A careful evaluation of the different development patterns that occurred in Italy, for example, might be of considerable use to third world planners. The debate on the urban bias (Lipton 1977) has often been carried on in too general terms. Instead, we need accurate studies on the flow of resources from countryside to cities (and within cities) and an assessment of the extent to which these flows overtax rural environments. This subject is very pertinent for countries like India and China.

Brazil and several other third world countries still blessed with an extensive economic frontier have the possibility of slowing down the urbanization process by properly choosing rural development strategies for frontier areas.

Third world countries are thus confronted by the twin problems of new rural-urban configurations and greater urban self-reliance. The former is predicated on egalitarian land tenure patterns and a symbiotic relation between a prosperous countryside and the small towns capable of servicing it. The latter implies community-based, need-oriented, and resource-conserving urban management styles, made possible by urban reforms giving fairer access to land and resources to low-income residents.

An even greater involvement of the state is necessary to implement development strategies based on the kind of IFES described above. Insofar as self-reliance means initiatives taken locally, it calls for a new paradigm of public policies empowering people to act and to take advantage of the diversity of ecological and cultural settings instead of supplying uniform, ready-made "solutions". Experimentation under real-life conditions should be encouraged both with respect to technologies and to the forms of organization of human endeavour.

Decentralized industrialization offers many advantages but the mistake of taking agriculture for a "bargain sector" (Chakravarty 1987) must not be repeated. Nor is it feasible to rely exclusively on market forces, even though integrated systems and agro-industries offer excellent opportunities for innovative private entrepreneurship.

To summarize, the prospect of rural industrialization with special reference to the application of biotechnology to biomass offers a convenient entry point into several important development strategy concerns:

- an alternative industrialization strategy that is resource-conserving, ecologically sustainable, fairly labour-intensive and yet economically viable thanks to the natural comparative advantage of tropical countries as far as primary biological productivity is concerned;
- a science and technology policy geared to the needs of such alternative industrialization;
- planning and project design methodologies emphasizing the systems approach and the need to seek ecosystem-specific and even site-specific solutions optimizing water, soil, energy, and technology management. This must be subjected to the constraints of social equity, ecological sustainability, economic viability, and balanced human settlement strategies, overcoming the traditional dichotomy of rural-agriculture/urban-industry, emphasizing the integration of local economies and seeking a symbiotic relationship between the countryside, small towns, and regional centres;
- attempting a synthesis between Ghandhi's concern for a human-centred economy and the respect of nature with Nehru's faith in science and technology put at the service of society and subordinated to a long-term national project. Perhaps the central question of our times is the underdevelopment of the political capability to manage technological change in accordance with democratically chosen social objectives instead of letting it take the lead.

How, then, can one conciliate the just distribution of employment and revenue with technical progress that eliminates jobs? How can one avoid in such cases the fragmentization of society and the phenomenon of marginalization? If industrialized countries have problems in developing constructive responses to these questions (and in investigating new cultural models), the situation of third world countries is even more compromised by demographic pressures, the backlog of unemployment, and the many external and internal constraints that inhibit the acceleration of growth.

Producing while Conserving

In countries rich in land and other resources, there is a considerable temptation to promote extensive growth based on the predatory incorporation of nature's capital in the current account. But this is done at the expense of long-term natural resource management.

With all too many third world countries, Brazil shares a schizophrenic tradition in this respect: there is rhetorical glorification of nature on the one hand, but its complete ransacking on the other, including the massive deforestation that is taking place today.

Deforestation is far from being the only ecological cost of uncontrolled growth. The industrial cesspool of Cubatao is another vivid manifestation of such problems. Rather than trying to list all of these "wastes of progress", let us try to determine the basic causes.

If left to their own devices, companies have a tendency to internalize their profits and externalize their costs, whether they be economic, social, or ecological. As Furtado (1988) pointed out, the unquantified costs of corporate decision-makers are particularly high in countries where capitalism has developed only recently. The heterogeneous social structure and the enormous built-in labour surplus favours a marked gap between micro- and macroeconomic productivity criteria (or more fittingly, micro-economic and macrosocial).

This gap is the result of market forces. The negative social consequences need to be corrected by regulatory action of the political system. Such measures are all the more urgent given the fact that the debt crisis encourages the over-exploitation of soils and other natural resources, favouring more than ever immediate short-term interests over long-term ones.

The internalization of ecological costs in the pricing system poses major problems and, in any case, would not be enough in itself. Environmental factors must be considered through a host of administrative measures and also through a redefinition of planning methods. A narrow path must be woven between uncontrolled economics and unabashed environmentalism.

The management of development implies an appropriate consideration of the means as much as the ends, of the ways as much as the rhythms of economic growth. It is the antithesis of the three-tiered reductionism that reduces development to economics, economics to growth, and growth to "productive" investments.

Rather than concentrating on rapid growth, which could even result in impoverishing a large part of the population, it would be better to develop at a lower social and economic cost. This *is* possible if existing production methods are better managed and better maintained, and if the omnipresent waste is reduced by recycling and making better use of resources. In other words, by combining the frugality of rural societies with modern scientific and technical knowledge.

This implies a development process that incorporates a holistic and horizontal vision of socio-ecosystems where the keyword is complementarity. It substitutes for the vertical and sectoral organization of the economy based on an increasingly narrow specialization. The systematic aspect merits attention. The natural ecosystem constitutes a paradigm that needs to be imitated by production systems designed accordingly.

For an Acceptable Adjustment

The debate on structural adjustment, imposed on indebted third world countries by international financial organizations, has bogged down with both sides refusing to move. More than 100 debt management plans have been proposed but none of them have been acceptable to both parties. The banks refuse to assume part of the substantial adjustment cost even though they made substantial profits from high interest charges with loans at variable rates. As for the debtor countries, they cannot pay the costs by themselves, especially in view of the fact that the debt crisis coincides with the fall of raw material prices, the deterioration of exchange rates, and the continuing refusal of industrialized countries to dismantle the trade barriers that prevent third world products from flooding their markets.

The calls in favour of a more "humane" adjustment, proposed by UNICEF, are certainly inspired by a legitimate concern. But this is hardly feasible as long as the hard core of adjustment policies consists of a reduction of public sector expenditures and the application of a set of conventional measures or orthodox financing which results in a lowering of production, consumption, and purchasing power. Employment suffers because of cuts involving investments whereas social conditions deteriorate if, on the contrary, the reductions affect first of all social programmes – subsidized housing, health, food, and education. The basic burden of readjustment falls upon those least able to bear it.

Between 1979 and 1983, expenditures on education in Latin America declined by 65 per cent, those on health in Africa by 56 per cent. On both continents, budgets allocated for housing, infrastructure, and urban transportation were cut by more than half. In citing such data, a recent World Bank publication[1] acknowledges that long-term growth, which is supposed to result from adjustment policies, will not by itself be able to eliminate absolute poverty and that it is, in addition, necessary to protect the poorer classes during the readjustment periods. At the conceptual level, this is a step in the right direction. But the most important steps remain to be taken: identifying the measures needed and testing their feasibility and efficiency.

Is it possible to re-establish the macro-economic balance without sacrificing too much economic growth? An input of additional resources would enable this, but as we have already seen, capital flows are defying good sense by flowing from the South to the North and nothing indicates a quick turnaround of the situation. That is why, in addition to the mobilization of additional resources, there is the need to use existing resources more efficiently by identifying all the forms of waste and proposing ways to reduce or eliminate them.

A comment on the concept of wastefulness in the use of resources is called for here. Resources may be used in excess of what is required by a given technology because of neglect, lack of skills, or ostentation. Then comes the wrong choice of technology, product, or the siting of production with respect to the market. Energy and other resources, for example, are being wasted in un-necessarily long inter-urban transportation lines. The same may happen with intra-urban commuting: energy, public investment funds, and people's time and money go into overcoming the excessive distance between the workplace and home.

Another instance of wastefulness is that of foregoing the opportunity to re-cover energy and materials from waste. At the other end of the spectrum is the much-discussed and subjective issue of assessing lifestyles from the viewpoint of their energy and material intensity.

Moving to human resources and wastefulness, one can identify a number of situations characterized by a disregard for people's willingness and ability to work, be it in the labour market, through self-employment, in the household sector, or by voluntarily engaging in unpaid social activity. In many ways, the waste of human potential is the worst form of wastefulness: such foregone opportunities are lost forever as human lives cannot be stocked for later use.

In effect, waste reduction amounts in macro-economic terms to the release of resources for development. It will be economically advantageous as long as its implementation costs do not exceed the costs of producing additional re-sources, with its net contribution to development financing being the differ-ence between these two costs. The economic advantage often goes hand in hand with environmental benefits.

A Development Reserve

The challenge is thus to find the right ways to transform these general approaches into policies that can result in concrete projects and programmes. In other words, to pass from the macro to the micro. Even in a planned economy, such as that which Kalecki (1978) was considering in his work, this transformation is fraught with problems.

The stakes, however, are significant. If an operating margin of 3 to 5 per cent of the GNP could be released over the next few years through the identification and reduction of waste, the exploitation of this "development reserve" could finally enable increased investment in social programmes more in keeping with the enormous needs.

Is this too optimistic a vision? Should the problem even be posed in terms of optimism or pessimism? On the contrary, one must emphasize the enormous moral and political responsibility of our generation. It is up to us to take this chance of profiting from or turning our backs on this opportunity.

The following section analyses the "development reserve" in Brazil, starting with flows of production, distribution, and consumption of goods and services. It does not, however, challenge the "objective function" or issue of consumption structures and lifestyles. This is not because it is unimportant; quite the contrary. But the choice of a social project is so laden ideologically and susceptible to controversy that it risks leading away from the subject at hand, which deals more modestly at the level of a greater instrumental efficiency – both economic and ecological. It can be said, however, that a social project based less on the "civilization of having" and more on the self-limitation of material desires would certainly result in a much larger "development reserve" while also generating considerable free time for non-economic activities.

Wastes That Could Be Reduced

The word "waste" covers a wide range of situations characterized by different degrees of over-use of resources and under-use of products. While certain forms of waste can be measured quite accurately, the evaluation of others is based on more subjective criteria.

There are relatively good data as far as post-harvest losses are concerned, notably cereals and other agricultural products that have been harvested from the fields but not yet served on someone's plate. The losses here are due to inadequate transport and storage as well as inadequate capacity of the agro-industry to process production surpluses. It appears feasible to try and "save" 10 per cent of agricultural production, which amounts to about 1 per cent of the GNP in Brazil.

Recycling is another sector that deserves serious attention. It is estimated that the 180 largest cities in Brazil, for example, produce about 41 tonnes per day of garbage. From this, it should be possible to extract each year about 5.1 million tonnes of compost, 790,000 tonnes of recyclable plastic wastes, 2 million tonnes of paper and cardboard, 450,000 tonnes of glass and 500,000 tonnes of various metals. If a market can be found for the compost, this would represent a total value of about US$500 million, or about 0.15 per cent of the GNP (Sachs 1989).

Another activity being developed is the exchange of industrial wastes. The Sao Paulo "waste exchange", for example, is designed to recycle 20 per cent of the industrial wastes of 70,000 industries in this state. Other states are following in its footsteps and there are also several sectoral waste exchanges, notably in the chemical industry.[2]

Progress in this field depends less on techniques than on the organization of separating the paper, metal and glass before it is collected. This sector represents a potential gold mine for environmental organizations, neighbourhood groups, and schools. The Shanghai Company for Resource Recuperation and Use manages 580 collection centres throughout the city and commercializes 30 types of material. Most of its 37,000 employees are engaged in recycling industrial wastes.[3]

On the other hand, innovative techniques are needed for the valorization of liquid wastes. Many efforts were made in the 19th century to use grey water produced by cities for agriculture, notably in Paris, Edinburgh, Berlin, and Milan. But little was then known about how to treat effluents. Ironically, now that efficient methods have been developed, the fact that wastes are misplaced resources seems to have been forgotten. The goal of 0.5 per cent of the GNP for the recuperation of urban and industrial wastes appears to be certainly within the realm of the possible.

Leaving aside other causes of underutilization of current production systems, let us consider another form of waste — the overconsumption of energy and raw materials in relation to the technical capacities of existing equipment. The oil crises of 1973 and 1979 brought us to the threshold of a new age of energy efficiency. Most industrialized countries have improved their energy efficiency by 15 to 30 per cent through energy conservation policies and technical progress, although it is true that structural changes in production systems and growth in the tertiary sector also contributed to this.

Despite such progress, the potential for energy conservation still remains enormous. Prototypes for new refrigerators consume 87 per cent less electricity than the current average, and the corresponding figure for air conditioners and electric water heaters is 75 per cent (Flavin and Dunning 1988:20).

According to Brown et al. (1988), efficiency gains of about 50 per cent could be realized in all sectors of the economy. In the construction industry, the concept of "intelligent buildings" enables one to project even more spectacular results. In the automobile industry, fleet performances by the end of the century should be 51 to 78 miles per gallon compared to 25 to 33 miles per gallon today.

But larger fuel savings will require major shifts of traffic from the private car to public transportation (four times more efficient per passenger mile), and from trucks to rail and water transport (three times more efficient per km tonne). Goldemberg et al. (1985) have shown that it is technically possible by the year 2020 to double the GNP per capita of industrial countries by reducing the installed capacity from 4.9 to 2.5 W, and to multiply per capita GNP of third world residents with an energy consumption per capita practically unchanged from current levels (1 W instead of 0.9 W).

Without tampering with structural variables, such as land-use planning, shortening of supply lines, substitution of rail and water transport for trucking, and public transportation for private cars, one could estimate that 0.25 to 0.5 per cent of the GNP could be saved through energy conservation in the transportation sector. The total of all energy conservation measures could thus reach, if not exceed, 1 per cent of the GNP in Brazil.

This inventory of the components of the "development reserve" could be completed by addressing the problem of resource conservation through better maintenance of equipment, infrastructure, buildings, and transportation systems. According to the International Labour Organization, poor maintenance is costing third world countries more than US$100 billion per year, and maybe even US$200 billion![4]

This is another gold mine waiting to be exploited. To what extent can third world countries profit from it? How many dozens, if not hundreds of thousands of self-financing jobs could thus be created? There are few data at hand, but 1 per cent of the GNP would appear to be a conservative estimate.

Through reducing waste, the "development reserve" could thus amount to about 4 to 5 per cent of the GNP, which would provide an operating margin for investment and social programmes that slightly exceeds the foregone gains constituted by the servicing of the debt. To this is added the underutilization of abundant resources, which is, strictly speaking, not a waste of resources but rather a development opportunity that is lost.

7

The Challenge of Biotechnology

A Pandora's Box

The second Green Revolution is coming of age. Biotechnology applied to plant and animal production and processing opens a "Pandora's box" of technological options in terms of increased yields and an ever wider variety of productive processes and end-products: foodstuffs, bio-energy, bio-plastics and other chemicals, pharmaceuticals, pulp and paper, etc.

For the first time in history human beings are sharing the responsibility of creating entirely new – and sometimes dangerous – forms of life. The challenging ethical problems posed by this situation will not be discussed in this chapter, which is concerned with the more pedestrian questions of policy options that can ensure that biotechnologies are used for the greater benefit of the rural and urban poor.

Under what conditions may biotechnologies become a lever of sustainable development by developing regenerative agriculture, agroforestry, and aquacultule instead of temporarily achieving sustainability by means of ever increasing inputs of commercial energy and nutrients?

How to plan for intensive use of renewable resources leading to more balanced rural-urban configurations and fairer employment and income distribution patterns by redeploying modern, small-scale bio-industries integrated into ecosystem-specific production systems that recycle and minimize the wastes? How can biotechnologies be used to improve the livelihood entitlements of poor peasants by making a greater range of technologies available to them that could ensure higher and more reliable yields on marginal lands with little water?

South-South Co-operation

The context in which biotechnologies have been developed to date does not lend itself to optimism. In third world countries they are rightly perceived as a menace because of their potential to enable industrialized countries to reduce their imports of tropical products. Numerous natural products which account for a significant share of third world exports are being displaced by biosynthethic substitutes or else by "tropical" products that have been developed to grow in a temperate climate. This ominous trend adds one more difficulty to the already depressed situation of the exporters of tropical commodities, even though in some cases new opportunities may also appear for third world producers.

Furthermore, biotechnologies come to the market almost entirely privatized, protected by patents and, to a great extent, controlled by powerful transnationals quick to buy the results of pioneering research and development performed by smaller, more creative enterprises. This degree of knowledge privatization considerably exceeds that observed during the first Green Revolution. The situation is diametrically opposed to the vision of science and technology becoming increasingly part of the "common heritage of humankind": biotechnologies are being commoditized to the disadvantage of third world countries and especially to their masses of poor peasants.

This explains the pessimistic conclusion about biotechnology by Costa Rica's Minister for Science and Technology, Rodrigo Zeledon:

Everything seems to indicate that the new technologies, far from being the instruments that will automatically save us from calamities, will simply serve to create new mechanisms for even more dependence. (Inter Press Service [Rome], Report 27, January 1988)

The political implication is clear. Third world countries should press international organizations to assist them in improving their access to the accumulated capital of knowledge, even though the prospect for advancing in this direction will remain dismal so long as commoditization and privatization are considered as panaceas for all economic ills.

More importantly, third world countries should develop their own research capability both for putting biotechnolgy to good use without depending on transnationals and for strengthening their position for the inevitable bargaining on the international technology market. The present unfavourable balance of power should not lead them to underestimate the paramount importance of biotechnologies for tropical countries, insofar as they can provide opportunities for a more efficient and intensive use of biomass produced both on the land and in the water.

On the contrary, third world countries need to assess the potential of an industrialization strategy based on the self-reliant but by no means autarkik development of biotechnologies, particularly those using biomass, "the fuel of development" (Hall and Overend 1987). Self-reliance here means autonomous decision-making with respect to research priorities, selective purchase of foreign technology, and promotion of collective self-reliance among third world countries.

South-South co-operation among countries that share similar natural conditions is urgently called for, going beyond the usual regional boundaries. The United Nations system should play a much more active role in bringing together scholars from Asia, Africa, and Latin America, arranging for the exchange of students, and promoting technological co-operation and joint industrial ventures. All of these facilities exist on paper, but in practice North-South circuits play a dominant if not exclusive role.

Paradoxically, developing countries are well placed to seize these new opportunities for spatially redeploying their secondary and even tertiary activities insofar as they are starting from scratch and need not struggle to overcome the inertia of the existing industrial structure. Indeed, leapfrogging should be seriously considered in this realm given the enormous savings arising from foregoing the heavy infrastructural investment that, sooner or later, large metropolitan centres will claim. The more so that countries like India already enjoy a considerable comparative advantage in qualified labour-intensive services such as software production or applied research required to accomplish this kind of strategy.

Considerable scope seems to exist to design decentralized, agro-energo-industrial systems not only in conformity with the ecological paradigm but also in such a way as to make maximum use of "combined technologies" blending along the production chain technologies of different vintages and capital intensity.

Co-operatives might offer a suitable institutional solution, allowing for the creation of strong and efficient enterprises without excessive concentration of capital in private hands. More generally, the potential of the social "third sector", as distinct from public and private ownership, should be explored in the search for new forms of partnership between the market, the state, and the civil sector within the paradigm of the "mixed economy".

The example of northeastern Italy offers a double lesson in this respect. It shows that it is possible to reach a very high level of industrialization and prosperity through the development of small-scale enterprises. But it also points to the elaborate social fabric and local policies which were conducive to such development: it would be a total mistake to credit this outcome to the free interplay of market forces.

Choosing Priorities

For Ducos and Joly (1988), the pace of discoveries and innovations in biotechnology is so rapid that most of the products that will reach the shelves of supermarkets by the end of the century are not yet known. While this vision may be exaggerated, there is no doubt about the amazing versatility of biotechnologies.

Much will depend, therefore, on the objectives set for the researchers. Powerful economic interests are likely to play a determining role in the choice of priorities: pesticide-resistant seeds instead of pest-resistant ones, fancy processed milk products catering to the urban middle class instead of a range of products maximizing the food intake per unit of currency spent on them, and so on. Finally, the considerable sophistication of the new techniques and the high capital outlays involved are likely to strengthen the position of large farms and businesses at the expense of smaller ones.

In other words, if left uncorrected, the present practices will reinforce the model of growth through inequality and of lopsided modernization, widening the gap between the élite and the growing mass of marginalized rural and urban poor.

Brazil stands as a striking example of the potential and limits of such growth through "maldevelopment". It managed to sustain an average 7 per cent annual rate of growth or GNP for 40 years – from 1940 to 1980 – and to build an integrated and modern industrial sector, but at a staggering social and ecological cost, which is seriously threatening its future.

Fortunately, Brazil and most tropical countries are blessed with climates that enable a high level of primary productivity of biomass, subject to the availability of land and water. Biotechnology can be directed to partly overcome land and water constraints and to greatly enhance productivity in areas with an adequate mix of such resources. Moreover, they can be instrumental in producing from biomass an array of useful products – food, fodder, energy, fertilizers, and a rapidly growing variety of intermediate and finished goods. Agro-industries based on the more efficient capturing of solar energy through plants are "sunrise" industries, literally and metaphorically.

Applications of biotechnology need to be pursued simultaneously at three levels: overcoming local constraints in environmental resource endowment, enhancing vegetal and animal biomass productivity, and industrial processing of biomass. Their joint effect may result in a permanent comparative advantage of tropical countries and ultimately unfold into an original pattern of decentralized industrialization. Such a pattern would be less capital-intensive than the one prevailing at present because of lower expenditure on large-scale urban infrastructures.

Properly planned and administered, biotechnology could generate considerable employment related to the management of soils, water, and forests for the production of biomass. Many biotechnology-based agro-industries are being located in rural areas and can adjust their rhythms to the seasonal variations of demand for agricultural labour. Their efficiency depends less on the scale of production than that of other industries.

At any rate, by operating a network of "agricultural refineries" that transform bulky biomass into less voluminous semi-products, considerable flexibility is obtained for the choice of the size and location of second level agro-industries.

Even dispersed and small industries can now interact with the national and international markets without incurring high transportation costs.

Biotechnology and modern communication facilities thus work in the same direction: they undermine the concepts of economies of scale and of concentration inherited from another industrial age and force us to assess them carefully case by case. They offer the possibility of "diffuse" industrialization, performed by small, innovative businesses that are quick at penetrating foreign markets.

Self-reliance in biotechnology is an ambitious but not impossible goal which requires a sustained public policy in research and development with clearly set priorities that benefit from a critical assessment of the lessons learned from the first Green Revolution (Glaeser 1987).

8

Sustainable Development

A Normative View

Ecological security is now recognized as an important aspect of the governance of the planet, alongside peace and the reduction of poverty. The three objectives are closely interwoven. A challenging task facing the international community is to intensify the war against poverty while averting further disruption of global ecological balances, or as the Brundtland Report puts it, to satisfy the present needs of humankind without undermining the capacity to meet the needs of future generations (World Commission on Environment and Development 1987).

Poverty is both a cause and an effect of environmental destruction. In their struggle to survive, the rural poor are forced to live from hand to mouth. As pointed out by Idriss Jazairi (1989), President of the International Fund for Agricultural Development (IFAD), they are caught in a self-destructive trap in which their immediate survival depends on over-exploitation of fragile resources. Population growth, ill-conceived development strategies, increasing debt, declining terms of trade, and natural disasters provoke the overuse of productive soils, forests, and waters. Thus, reducing poverty is a direct way to ensure environmentally sound development.

The reverse loop is illustrated by the plight of the urban poor in Mexico City. Their poverty is made worse by the degraded environmental and health conditions prevailing in this city on the brink of ecological disaster. The situation in many other large Latin American cities is also deteriorating under the combined effect of inadequate housing, deficient sanitation, excessive concentration of polluting industries, and the dominance of individual cars in urban transportation.

The way out from the double catch of poverty and environmental disruption calls for more economic growth while changing drastically its forms, contents, and social uses. The normative concept of sustainable development, as presented in the Brundtland Report, reflects this double preoccupation incorporating almost two decades of a worldwide debate. Started at the Founex seminar in 1971, this debate was punctuated by several international conferences (notably those in Stockholm and Vancouver), enriched by the work of UNESCO, UNEP, and other specialized agencies, and nurtured by the reports sponsored by the Club of Rome and the manifold contributions of environmental organizations.

The question before us now is how to speed up genuine socio-economic development through engaging in a positive sum game with nature instead of continuing the predatory practices that deplete at an alarming rate the capital of nature and undermine the life-support systems. How can we replace the concept of domination of nature, central to our technological civilization, with one of symbiosis between society and nature?

Sustainability is a dynamic concept that takes into consideration the expanding needs of a growing world population, thus implying steady growth. It encompasses the new awareness of the limits of "Spaceship Earth" and of the fragility of its global ecological balances, a need-oriented approach to socio-economic development, and the recognition of the fundamental role of cultural autonomy. It has a double function: the direction in which to move and a set of criteria to evaluate more specific actions.

Time for Action

It is easy to dismiss sustainable development as one more utopia, arguing that entrenched vested interests press for more of the same, that is, savage growth and imitative modernization; that environmental protection is expensive and, therefore, should wait for better times; and that the present economic and social situation in many third world countries is too critical for implementing any such changes.

The conflicting interests around the choice of a development strategy should not be minimized, but neither should they lead to the circular argument that difficult political choices are not eligible for discussion because they are politically difficult! Priority for sustainable development cannot be argued on the grounds of narrow micro-economic calculation. It calls for a political decision based on a long-term view of the country's interest, its share of responsibility in the global management of the planet, and an appreciation of the positive externalities created for the population at large by arresting further environmental degradation.

This being said, the additional costs of sustainable development, as compared with the business-as-usual scenario, pose a problem in market economies. Companies resist the internalization of costs that up to now have been externalized and they seek public subsidies to cope with stricter environmental regulations. Hence the publicity given to those cases when the enforcement of such regulations imposes a real financial burden. In contrast, the successful instances of profitable shifts to low-waste technologies are publicized much less for obvious reasons. Yet the scope for a positive sum game, in which economic and environmental gains go hand in hand, is far from negligible and can even be increased through well directed research and experimentation.

Furthermore, what appears in the short term as a trade-off between an environmental gain and more growth often amounts to a choice between a preventative action today and a much more expensive remedial action tomorrow (and therefore to a trade-off between the present and the future, a dilemma constantly faced by planners).

Finally, the conflict between environmental and economic objectives disappears in all the actions directed at resource conservation, reduction of wastefulness, recycling, and maintenance of vehicles, equipment, and infrastructure. In macro-economic terms, the resources thus saved constitute a potential source of development, not to speak of the employment thus generated that pays for itself.

Scientific evidence increasingly points to the dangers involved in the Faustian bargain: our careless use of ever mightier technologies on a scale which now compromises global ecological balances. The time has come to redress the wrong done to the planet in order to preserve a niche for future generations. As already argued, global change must be addressed by a multiplicity of local actions with the main responsibility resting with industrialized countries. As for third world countries, the present crisis should be viewed as an opportunity to leap-frog into a sustainable development model.

Instead of reproducing the techno-structures of the industrial countries, third world countries should attempt to evolve new agro-silvicultural industrial patterns and rural-urban configurations by making better use of their renewable resources. In this way, they could transform into a permanent comparative advantage the primary biological productivity of their terrestial and aquatic tropical ecosystems.

An International Action Plan

There are a variety of ways in which the international development community and lending institutions could assist third world countries in their transition towards sustainable development. Some possible lines of actions are discussed below.

Whatever the speed of the transition towards sustainable development, the backlog of unattended environmental demands and damage already done by ill-conceived development projects is such that a substantial increase in the volume of resources available for conventional environmental projects is needed.

Environmental management of the "global commons" poses difficult institutional and financial problems. Yet sustainable development often depends on the good environmental health of the life-support system that constitutes the common heritage of humanity. A decisive improvement in this respect could be obtained by establishing an international environment fund financed by token user fees (Silk 1989) that could be used for implementing sustainable development strategies with special emphasis on energy conservation and reforestation programmes in both industrialized and developing countries.

Providing additional funding for environmental projects does not pose complex conceptual problems. In contrast, internalizing the environmental dimension in all ongoing and future development projects calls for a serious methodological effort. Environmental impact evaluation procedures are still far from satisfactory and many remain quite cumbersome despite the recent advances accomplished in this field by some international lending agencies. The project-by-project approach isolated from the broader context makes it difficult, if not impossible, however, to consider second- and third-level impacts and the cumulative and often irreversible processes occurring downstream. Many qualitative and social aspects are left out, and the purely formal participation of the people concerned in the assessment is largely ineffective.

Further methodological work should explore the following four directions.

1. Differentiating the approaches depending on the size and type of the project, giving preference to comprehensive programmes rather than collections of individual projects.

2. For large hydro-electrical, mining, and transportation projects and for other comprehensive programmes, instituting as a rule their evaluation in the context of explicit regional development strategies or plans.

All development planners should contemplate the lesson of the unprecedented ecological disaster now unfolding around the Aral Sea in the Soviet Union. It is the consequence of a series of upstream irrigation projects which diverted too much water from the Amu Darya and Sir Darya rivers. Each of these projects taken in isolation may have had a favourable cost-benefit ratio in economic terms. Yet their cumulative impact has been severely underestimated, ultimately resulting in a tragedy.

The Aral Sea has shrunk by two-fifths since 1960, leaving behind over 20,000 sq km of salty, man-made desert contaminated by agrochemicals. This has victimized over 3 million people and the rice and cotton crops for which these sacrifices were made are now showing signs of ecological strain themselves.

The Great Carajas project in Brazil is another case in point. It is being implemented as a succession of individual mining, transportation, industrial, agro-pastoral, and forestry projects (including the controversial use of charcoal for the production of cast iron). Its critics point out that their cumulative ecological effects are not being properly assessed and that it would have been much wiser to produce first a regional development strategy for the whole East Amazon.

3. Improving the analytical tools and procedures used in the evaluation by interrelating social, economic, and ecological indicators so as to arrive at a satisfactory approximation of a "sustainable development conditionality" more acceptable to developing countries than the versions of conditionality now used by international lending agencies and donor countries.

In parallel, a systematic effort associating the member countries should aim at identifying the appropriate policy instruments and packages to implement sustainable development strategies in the context of "mixed economies". A better knowledge of the range of possibilities in this respect and of actual experiences – both positive and negative – is urgently needed.

The question of why governments and the international community have been so slow to implement the consensual resolutions arrived at in Stockholm almost 20 years ago must also be addressed. In order to understand and overcome these obstacles, more needs to be known about the functioning of the "conservative dynamism" of the vested interests and the "bureaucratic rings" that link private and public business.

4. Insofar as sustainable development strategies must rely on a multiplicity of site-specific projects, genuine – not merely rhetorical – participation of the people concerned through, inter alia, representative citizens organizations is required at all stages of programme and project formulation, evaluation, and implementation.

This, however, poses difficult institutional problems. How to establish an effective dialogue between all the actors involved in grassroots projects? How to protect the non-dominant players from the pressures of the dominant ones (business and state)? How to identify the legitimacy of citizens associations in a pluralistic setup? How to enable them to take an effective part in project formulation and evaluation by giving them access to sources of technical expertise? How important for these grassroots organizations is access to sources of financing other than public subsidies? What kind of "banks for the poor" are required? What is the scope for the "social sector" (co-operatives, mutual aid, and non-profit associations) in the "mixed economy"?

Innovative Resource-use Patterns

While environmental impact assessments can stop environmentally disruptive projects or modify those that can be improved at some additional cost, they are essentially defensive techniques. Looking for new, environmentally sustainable, socially useful, economically efficient resource-use and management patterns is, in contrast, an approach aimed at exploring and expanding the space of positive sum games with nature. This could best be achieved through life-size experiments of integrated production systems adapted to the diverse ecosystems (arid- and semi-arid regions, tropical rain forests, highlands, estuaries,

and aquatic ecosystems) and responding to the ecodevelopment criteria introduced above:
- recognized social utility of the product mix and equitable distribution of the income generated, which in turn calls for managing the technological pluralism by resorting to "combined technologies", that is, making a catalytic use of capital-intensive techniques in some links of the production chain while privileging the labour-intensive ones in the remainder;
- ecological sustainability of production with special emphasis on regenerative agriculture, agro-forestry, and aquaculture. Whenever possible, these primary activities should be combined with the production of biomass energy and an array of biomass processing industries (food and non-food);
- economic efficiency, the aim being the intensification of resource use without impairing the long-term productivity of the life-support systems.

A good starting point for the design of modern integrated systems is provided by the analysis of traditional resource management patterns, such as that done for the Spanish dehesa by Perez (1986). Priority should be given to agro-forestry systems for tropical rain forests, given the urgent need to stop deforestation and the considerable time-lags required to develop suitable solutions for the millions of hectares that will need to be planted.

Rapid action is also required to arrest the destruction of mangroves and fragile coastal ecosystems. The best response would be to develop aquaculture-based production systems, the more so that the neolithic revolution has not yet been completed: hunting and gathering are still the dominant techniques used with respect to aquatic biological resources.

The shift to cultivation of aquatic resources could bring about a quantum jump. Many third world countries enjoy easy access to the sea as well as being well endowed with biologically fertile estuaries and lagoons, and inland water systems. Fish ponds are the least expensive method of reclaiming degraded agricultural land, at least in coastal areas, and large-scale production of solar dried fish could greatly improve the diet of the urban poor. Furthermore, the research required to sustain aquaculture programmes is less expensive than in many other sectors of the economy.

The fundamental importance of innovative thinking regarding resource-use patterns capable of harmonizing social, economic, and ecological concerns – the very essence of sustainable development – cannot be overemphasized.

9

Conclusion

The Future Remains Open

The environment of 40 years from now will be what we make it. We have both the technology to blow the planet apart and the wisdom to cultivate it into a polycultural garden complete with wilderness areas. The carrying capacity of "Spaceship Earth" is not unlimited, but it is elastic. This elasticity depends on consumption models, spatial configurations, technology, and institutional and cultural factors in the broadest sense of these terms. Social and cultural limits must take precedence over physical limits, at least for the decades to come.

The dashed hopes of the Stockholm Conference demonstrate the danger of drift, in which attention is focused on environmental discourse rather than on the redefinition of development strategies; sensitivity towards nature often emerges only after it has been destroyed. Have we already reached this stage? Is the massacre to go on?

Whether we admit it or not, the responsibility of our generation is enormous (Brown Weiss 1989). We are shaping not only our future, but that of our children and their children (without even speaking of the other species with which we share this earth). We will be judged by our ability to question our development patterns (and ourselves) and thus our ability to break with the dominant model in the West, the East, and the South. What can we do in these circumstances, here and now?

First of all, we can clarify the stakes of development, the margins of freedom, and the constraints which exist locally in a variety of forms. Understanding these factors is clearly the first step towards action.

Our capacity for analysis, combining all factors — ecological, cultural, institutional, personal, and socio-economic — remains limited. For that reason, we must envisage a training effort tackling educational programmes at all levels and in all channels. Multi-disciplinarity is not achieved through the juxtaposition of narrow-minded specialists; it entails an open-mindedness to dialogue on the part of all individuals, professionals and citizens alike.

Prominent among the necessary educational aids is an ecological history of humanity, conceived as a systematic exploration of the ecosystem/culture inter-relationship and revolving around the themes of food production, housing, energy, and so on. Such a history would make it possible to assess the adaptability of a culture to the various natural environments or, to put it the other way around, to compare the ingenuity of the various cultures in overcoming constraints and seizing opportunities within a particular ecosystem. The concept of "resourcefulness" is at the heart of the development process.

While a sound knowledge of history will stimulate the failing social imagination (by giving it guidance and, at the same time, by identifying anti-models to be avoided), the concept of new production systems meeting the triple criteria of social equity, ecological sustainability, and economic viability will benefit from being viewed in terms of the natural ecosystem, emphasis being placed on the complementarities between the various productions.

The exploitation of natural resources under ecologically viable conditions (the economy of permanence, as Gandhi called it) is by far the most effective and durable form of environmental protection and elimination of "raubwirschaft". This is why the programmes described above warrant a priority that has so far been denied by international organizations.

Such organizations clearly have a responsibility to deal with the "international commons". But progress in this field, if any, is terribly slow despite the magnitude of the stakes. While the establishment of supranational bodies endowed with genuine decision-making and managerial power seems to be out of the question for the time being, there remains the method of agreement. It would be strengthened if international efforts could at least count on automatic financing. Various such proposals were made in the wake of the Stockholm Conference (Steinberg and Yager 1978), but there has been very little acceptance of this idea to date.

Yet only international taxation, however modest, would give the United Nations and other organizations the financial autonomy that is absolutely necessary to elude the pressures of the great powers who are also the principal donors. In the Middle Ages, the Church levied the tithe. The United Nations would certainly be content with a tithe of a *tithe of a tithe* of the gross world product, which now exceeds US$10 trillion. A world-wide tax of US$1 per million (graduated in such a way that the rich countries would pay more and the poor would be exempt) would bring in US$10 billion. This is ten times more than the current annual budget of the United Nations!

In concluding, it should be emphasized that good work is in progress but much more needs to be done. There is, in particular, a great need for more systematic analysis of patterns of resource use in diverse ecosystems and forms of human adaptation to given natural settings that can show the diverse ways in which people manage to overcome the constraints of their environment and identify opportunities for a better life.

Let us hope that the 1992 United Nations Conference on Environment and

Development will accelerate political and economic movement in this direction – as time is no longer on our side.

Each generation modifies its historical accounts, whether they be written or oral. Ours should be recording the ecological history of humanity in order to ensure its future.

Appendix I: FEN Programme Activities

Integrated Food-Energy Systems

The first FEN activity in this field was a very successful study and conference tour in 1983 by four Brazilian special UNU fellows to Senegal, France, India, and China.

This six-week tour had a double purpose:
– to visit research centres involved in the field of food and energy; and
– to participate in seminars and workshops arranged for the occasion by the host institutions in collaboration with the UNU.

As part of the tour, seminars were held at ENDA, Dakar; CESTA, Paris; NISTADS, New Delhi; CRESSIDA, Calcutta; the Institute of Energy Conversion, Guangzhou; and the Biogas Research Institute, Chengdu, China.

There were also many visits to other research centres and government agencies in the cities mentioned as well as in Bombay, Madras, and Beijing.

The mission ended with a visit to UNU headquarters in Tokyo and a presentation to senior staff members.

This exchange of views between third world scientists proved very enriching. While the need for developing integrated food-energy systems was felt everywhere in the countries visited, it was noted that different socio-cultural mentalities affect in no small way the manner in which such systems can be implemented. Another essential finding was the need for emphasizing the fundamental importance of local participation in such projects and for ensuring that a wide range of technological configurations is considered for their development.

As a follow-up to this study tour, the Brazilian fellows hosted numerous counterparts from Senegal, India, and China as well as from several lusophone countries in Africa to a series of workshops, conferences, and field trips organized by various institutions in Brazil with the additional support of UNDP (United Nations Development Programme) and UNESCO. Several FEN researchers also participated in some of these events.

The first regular year of activity for FEN was highlighted by two international conferences. In September 1984, a seminar on "Ecosystems, Food and Energy" was held in Brasilia under the joint auspices of the UNU, UNESCO, and the following Brazilian institutions: FINEP (Brazilian Agency for Research Financing), CNPq, EMBRAPA (Brazilian Agency for Agricultural Research), and the University of Brasilia. It was attended by more than 80 researchers from throughout Latin America and leading experts from other continents.

The principal outcome of the conference, whose closing session was presided over by Dr. Kinhide Mushakoji, UNU Vice-Rector for Regional and Global Studies, was the identification of inputs for the elaboration, evaluation, and installation of integrated food-energy projects as well as the development of appropriate research programmes.

Proceedings of this conference were published in 1986 by UNESCO in three volumes in their original language (mostly Portuguese) in addition to a volume of summaries in English that also includes an analytical report on the conference prepared by Dr. Ben Wisner.

Interest in integrated food-energy systems subsequently progressed from the research to the development phase, where it received attention at the educational and policy level. This was particularly true in Brazil, where the FEN-inspired research programme made this country the major showcase for integrated food-energy systems. The major activity in this area was the "Agro-Energy Communities" programme in Brazil funded by FINEP. Involving numerous research and development organizations in Brazil, it was the most comprehensive application of FEN concepts in this field.

Many Brazilian universities and research institutions continue to be actively involved in the development of integrated food-energy systems, as reflected in the attention given to this subject by the 4th Brazilian Congress of Energy and seminars held with the participation of the FEN programme director in Campinas, Porto Alegre, Brasilia, and Itabuna.

The FEN methodology highlighted at the UNU Brasilia conference in 1984 also found its place in the teaching curricula of many educational centres, including the University of Campinas, the University of Sao Paulo, and other research centres. It was also adopted at the new Centre for Agricultural Sciences and Environment in Itabuna.

This momentum was reflected in the decision of CENDEC (a training centre for economic development that reports directly to the planning secretariat of the president of Brazil) to organize a seminar in August 1988 on "Resource Use Patterns, Employment and Development Training", and to ask the programme director to provide the background paper.

A major follow-up to the 1984 Brasilia conference was the "Food-Energy Nexus and Ecosystem" conference held in February 1986 in new Delhi. It was organized by the Indian Institute of Management (IIM) with the additional support of the Indian Department of Non-conventional Energy Sources (DNES) and UNESCO. It compared the design and operation of existing integrated food-energy production systems developed in different countries under diverse ecological and socio-economic conditions.

One of the resulting recommendations was the establishment of a permanent working group to study FEN and third world development strategies. DNES subsequently announced that it would provide logistical suport for this group to be based at the Global Energy Centre in New Delhi. A comprehensive report on this conference was prepared by Dr. Ben Wisner and published as part of FEN's research report series.

After the New Delhi conference, Drs. T.K. Moulik (IIM, India) and Emilio La Rovere (FINEP, Brazil) jointly prepared in June 1987 a feasibility report for "Establishing a Permanent International Network on Biomass-based Agro-Industrial-Energy Systems" designed to advance South-South co-operation on integrated food-energy systems. This work will hopefully be undertaken with the support of the governments of both India (through the Global Energy Centre in New Delhi) and Brazil.

The third international conference on integrated food-energy systems was organized by CAST (Chinese Association for Science and Technology) on the basis of a comparative study of such systems in seven ecologically diverse regions of China. It took place in Changzhou in October with the participation of the programme director as well as FEN colleague, Dr. T.K. Moulik, of the IIM.

Preparatory work for a comparative study on "Ecosystems and Cultural Diversity" was done in collaboration with the Federal University of Alagoas in Brazil and the Marga Institute in Sri Lanka. It was designed to analyse resource-use patterns in order to identify how different cultures sharing similar ecological conditions have developed different methods of producing and using food and energy.

A consultant's mission to Managua in July 1985 analysed the feasibility of using organic wastes from that city's food markets for the production of animal fodder and concluded that there was considerable institutional support for using such wastes for this purpose as well as for the production of compost for use in kitchen gardens.

A mission to assess the potential of integrated food-energy systems to improver access to food and energy in Rwanda concluded that, apart from a few opportunities associated with commercial processing or institutional livestock operations, the improvement of composting methods at the family or "hill" level held the most promise in this field. Proposals to continue this work were submitted to the Government of Rwanda and IFAD (International Fund for Agricultural Development).

Another important development was the start in 1986 of collaboration with the Asian and Pacific Devlopment Centre (APDC). Six country reports reviewing integrated food-energy systems in Southeast Asia previously prepared by the APDC were discussed during a workshop held in conjunction with the New Delhi conference described above, and plans for continued co-operation were made. The subsequent preparation by the APDC of six bibliographies on integrated food-energy systems in this region represented a total of over 2,000 pages of documentation and analysis that was edited by the APDC for joint publication.

A research proposal was also developed in conjunction with the Biotechnology Programme of the UNU Development Studies Division to assess the social impact of new or foreign food and energy technologies and to determine the potential of biotechnology in improving traditional techniques in this field. The conceptual framework prepared by J.-P. Peemans of the Université Catholique de Louvain was published as a FEN research report.

Alternative Urban Development Strategies

Several research projects in this field were started in 1984. In Brazil, the Center for Study and Documentation of Community Action (CEDAC) began studying the experience in Osasco with community kitchens designed to reduce nutritional problems through the provision of communal cooking facilities. Its final report describes the difficulties encountered in continuing the community kitchens that it established in this suburb of Sao Paulo.

In Bombay, the Tata Institute of Social Sciences identified two poor neighbourhoods to involve in an experiment on urban agriculture and social organization. The report of this "Action Research Project", which was published as an occasional paper, deals *inter alia* with the aspirational obstacles facing rural migrants in urban areas who came to the city to find a job, considering farming a demeaning activity that they had left behind.

In Buenos Aires, the Urban and Regional Studies Centre (CEUR) began a research project on strategies to improve access to food and energy in urban areas of Argentina. An interim report (in Spanish) on "Alternative Strategies for Improving Food and Energy Availability in Argentine Cities" was prepared in addition to a final report (also in Spanish) on urban agriculture in Buenos Aires that was co-published with UNESCO. It describes a dozen case studies located throughout greater Buenos Aires that were analysed and compared with similar activities in other countries, both developed and third world. While the emphasis was on the role of urban gardening, other means of access to food were also considered, including food subsidies and supplements.

The CEUR authors concluded that significant social, environmental, and nutritional advantages accrue to families that garden, notably providing up to 30 per cent of their food needs. It was pointed out, however, that in the already huge and burgeoning third world cities, the impact of kitchen gardens on the overall nutritional situation would remain low in the near- and medium-term. The results of this research were also presented at the October 1986 International Council Meeting in Paris of UNESCO's Man and the Biosphere (MAB) programme. A second publication resulting from the CEUR research in Buenos Aires, entitled "Commercial Agriculture in Greater Buenos Aires: Experience and Perspectives", addressing green belts for commercial food production, was published in January 1987.

In September 1984, the City of Sao Paulo joined with ECLAC and the UNU in organizing a symposium on "Latin American Cities Facing a Crisis". A striking feature of this event was the fertility of the discussions, a result of the widely varied background of the participants, including mayors from several Latin American cities, who agreed to establish a follow-up network.

An important conclusion was the confirmation of certain room for manoeuvering in which to develop alternative strategies for urban development that can exploit the untapped or wasted physical and human resources of urban ecosystems. The proceedings of this conference were published in 1986 in Portuguese by the Sao Paulo Municipal Planning Secretariat (SEMPLA).

As a follow-up, the Sao Paulo Municipal Administration Research and Study Centre (CEPAM) established a communications network on urban innovations, known as RE-CEM. It prepared a report for FEN on "Municipal Participation in Local Food Production and Distribution" based on its data bank of over 1,200 urban projects. With the support of the Federal Savings Bank, the scope of this project was extended to use this network as a basis for exchanging such information throughout the entire country, with plans to cover Latin America and eventually third world countries on other continents through a South-South network.

In August 1986, an international workshop on FEN in Latin American cities was convened in Sao Paulo in collaboration with the Faculty of Architecture and Urbanism of Sao Paulo University. It was timed to coincide with MAB UNESCO's conference of "Latin American Cities: Looking Forty Years Ahead". The FEN meeting brought together the leaders of FEN urban projects in Latin America in order to review the results to date of their research on alternative urban development strategies and to define ways of further co-operation, including the common publication of their research results.

A major meeting in this field was held in 1985 at the National Research Council in Washington, when the Board on Science and Technology in International Development (BOSTID) and FEN brought together international representatives of networks on urban alternatives to discuss innovative ways of improving the use of urban resources. One of the main conclusions was the identification of a large communication gap between the sources of technical knowledge in this field and the different people who need such information, particularly at the local level.

A workshop on "Improving Access to Food and Energy for the Urban Deprived" was organized by ENDA (with additional support from the European Economic Community) in December 1985 in Addis Ababa. It brought together researchers from 11 countries (mostly African) and was for many their first exposure to an interdisciplinary analysis of these problems in an African context. Among the recommendations made was that the first need was for more information dissemination, particularly South-South and between local people actually working in the field. A report of the conference was prepared by ENDA and is available in French and English. As a result of this workshop, follow-up research in African cities was done by ENDA and the results were published in FEN's series of research reports.

The concepts and ideas emphasized in FEN research found their way into a joint UNU/MAB-UNESCO proposal on "Urban Ecosystems: Resource Use Patterns and Employment Generation", involving a comparative analysis of Buenos Aires, Sao Paulo, and Santiago. The MAB/Sao Paulo project (in conjunction with CETESB and the Agency for Application of Energy), CEUR, and the Catholic University of Santiago assumed responsibility for local funding. The Federal Secretariat for Environment (SEMA) also expressed interest in seeing Recife used as a case study for Brazil.

The first phase of a project on urban self-reliance undertaken by IFDA (International Foundation for Development Alternatives) resulted in the publication in 1985 of directory of 150 institutions and projects working in this field. Work continued throughout 1986 on this project and, as a result, an updated and expanded version of this global directory on institutions and projects involved in urban self-reliance was published in 1987.

Several new projects were started in 1985. Among them were "Food and Energy in Urban Areas: the Case of Caracas" with CIEDA (energy, development, and environment research centre) in Venezuela, which was facilitated by a FEN mission to Caracas. The main goals of this research were to diagnose food and energy consumption in low-income households and to propose policies to help improve access to these basic needs through increased self-reliance. A final report, in Spanish, was submitted.

A companion project entitled "Food, Energy and Public Services Self-reliance in Mexico City" was undertaken by PROCALLI (Support to Mexican Housing). A final report entitled "Research Project on Food, Energy and Public Services Self-reliance in Mexico City Metropolitan Area" explores the interconnections between energy systems and local development dynamics in Mexico. A complementary report by CECODES was also produced.

A similar project on "Household Energy Redesign and Food Production in Marginal Areas of Santiago" was undertaken by PRIEN (energy research centre, University of Chile) with a final report in Spanish. An interim report entitled "Energy and Improvement of Urban Nutrition: A Methodology for Action-oriented Research in the Local Setting" was prepared in English.

Also started in 1985 was a project with ECLAC (UN Economic Commission for Latin America and the Caribbean) on assessing innovative policies and strategies for alternative urban development styles in Latin America and the Caribbean. This study was an element of the work undertaken by the ECLAC/UNCHS (United Nations Centre for Human Settlements) Joint Unit on Human Settlements. It included training activities, technical co-operation, and research in the areas of metropolitan planning and management, design and management of plans and projects at the local level, community participation, and technologies for housing and for the provision of social services and infrastructure for the urban poor. A report entitled "Metropolization in Latin America and the Caribbean" was prepared.

Two case studies on "Equity, Efficiency and Sustainability in Urban Energy Development" were prepared by the Environmental Liaison Centre (ELC) in Nairobi for the cities of Nairobi and Ahmedabad respectively, and were presented at the joint ELC/IAS (Institute of Solar Architecture) international seminar held on this subject in La Plata, Argentina, which was supported by the Canadian International Development Agency (CIDA).

A new project started in 1986 was "Everyday Structures and the Working of the Real Economy in the City: Going Beyond the Formal/Informal Dichotomy". Co-ordinated by the College of Mexico, it was designed to prepare critical interdisciplinary surveys of the situation in the following cities: Cairo, Mexico, New Delhi, Rio de Janeiro, Rome, Santiago, and New York. An internatonal workshop for this project was held 17–19 June 1987 at the Maison des Sciences de l'Homme in Paris to review the surveys. As a result of this workshop, which brought together the authors and several reviewers, the papers were revised for publication in English, French, and Spanish. A more immediate spin-off was the publication of a special issue of the Latin American Planning Review, *SIAP*, dedicated to FEN urban research (see Sanchez 1988, Appendix II).

Another major event in 1987 was designed to capitalize on the International Year of Shelter for the Homeless. FEN accepted a proposal from GRET (Technological Research and Exchange Group) for, *inter alia*, the organization of a seminar on communication and urban strategies and the development of tools designed to strengthen and build upon the alternative urban development networks that emerged during 1987. This project also included working sessions at regional NGO workshops that were held in Sao Paulo and Nairobi as well as a special workshop at the Habitat Forum in Berlin. Two reports of this seminar, which was held June 22–26 at the Fondation des Sciences Politiques in Paris, were prepared: an Exective Summary (in English) and "Communication and Urban Self-reliance Strategies" (in French).

A project entitled "Production of Food in Big Urban Agglomerations: Poland as a Case Study" was completed by a team of researchers based at the University of Warsaw.

The purpose of this study was to analyse and evaluate Polish experience in the field of small-scale food production in urban areas with a view towards their relevance in the formulation of research needs and policy decisions in other countries. The final report was edited and published in FEN's series of research reports.

The last "official" FEN activity was an international conference entitled: "Cities⁻ What For?", sponsored by the City of Rennes, the Fondation Diderot, the League for Education, and the Fondation de la Maison des Sciences de l'Homme. It was held 25–27 November 1987 in Rennes, France.

Training, Networking, and Publications

The FEN programme formally started on 1 January 1983 and during this first year of activity it advanced from initial planning work to action on several fronts. Substantial groundwork was laid in terms of basic organization, professional contacts, conceptual development, and institutional arrangements, including the establishment of an administrative core unit and resource centre in Paris.

The UNU Global Food-Energy Modelling Project, directed by Dr. John Robinson, held its final session in June 1983 under the auspices of FEN at the Maison des Sciences de l'Homme, Paris. The results of this project were presented to an international audience including representatives of the FAO (Food and Agriculture Organization), IIASA (International Institute for Applied Systems Analysis), UNESCO, OECD (Organization for Economic Co-operation and Development), UNEP (UN Environment Programme), and the IIM as well as of universities in several countries. Following presentations by Professor Maurice Levy (who, as UNU programme director for energy studies, launched the study) and Alexander King, president of the Club of Rome, the project team described the results based on the use of three global models: SARUM, UNITAR, and UNITAD.

The first FEN newsletter was published in 1983 in conjunction with COPPE at the Federal University of Rio de Janeiro, FINEP, and EMBRAPA. It included an introductory article by the programme director and analyses of two integrated food-energy systems in Brazil in addition to news of other FEN activities.

The second issue of the newsletter appeared in 1985 and apart from a conceptual article by the programme director, it included an annotated bibliography and reports on various FEN activities in addition to articles on the study tour of Brazilian researchers, the "Agro-Energy Community" project, and a research project by Bologna on the self-reliant city.

The third issue of the newsletter was published in 1986 and, for the first time, a Spanish edition appeared in the *Boletin de medio ambiente y urbanizacion* published by CLACSO (Latin American Council on Social Sciences) and distributed to researchers throughout Latin America. It included a feature article on self-sustaining rural communities in Brazil in addition to reports on other past and planned FEN activities.

Four publications were published in 1984 in collaboration with the International Research Centre on Environment and Development (CIRED) and the French Agency for Energy Management (AFME). Please refer to Appendix II for the complete list of FEN publications.

Throughout the programme, a graduate seminar on FEN issues was held at the Ecole des Hautes Etudes en Sciences Sociales in Paris. Some of the students who participated in this seminar wrote papers inspired by FEN and several theses were prepared. Doctoral dissertations already completed include: "Systèmes alimenaires et rapports sociaux: structures du quotidien et production alimentaire dans le Catatumbo (Colombie)" by Dr. F. Pinton; "L'alimentation et l'énergie dans l'économie paysanne andine" by Dr. N. Herrera; and "L'enégie dans les systèmes de production" by Dr. S. Schilizzi.

Two UNU fellows arrived at the school in 1985. Discussions were also held with UNESCO and the Faculty of Architecture and Urbanism at Sao Paulo University to organize a workshop on teaching curricula and the role of universities in promoting self-reliant urban development strategies.

FEN urban research constituted a special issue of *Economie et Humanisme* (no. 282, mars-avril, 1985), was featured twice in *Development Forum* (February/March and July/ August 1985), was mentioned frequently in *Ecodevelopment News*, and was also covered by more popular media, both print and electronic.

A special issue of *Development* (no. 4 [1986]) was devoted to the urban self-reliance theme. Apart from the opening article by the programme director, entitled "Work, Food and Energy in Urban Development", it included contributions by the following FEN researchers: Pablo Gutman, Janice Perlman, Céline Sachs, and David Morris.

The October 1986 issue of the UNU newsletter, "*Work in Progress*, 10(1), was dedicated to FEN activities. Published in English, French, Spanish, and Japanese, it served as an excellent summary of the issues addressed by FEN and its widespread circulation resulted in numerous requests for FEN publications.

A special issue of the UNU *Food and Nutrition Bulletin*, 9(2) prepared by FEN and published in June 1987, included articles on urban agriculture by FEN consultants, including Pablo Gutman, Isabel Wade, Yeu-man Yeung, and Jerzy Kleer.

A total of 27 research reports and two occasional papers were published as part of FEN's publication programme. These covered both the rural and urban aspects of FEN in many countries around the world and proved to be a very cost-effective way of encouraging and disseminating the results of such research.

In conjunction with the recommendation of the African ministers of environment to establish 180 ecodevelopment projects to help ensure food and energy self-sufficiency, FEN was also contacted by UNEP to provide teaching and background material for this ambitious programme.

Finally, the Foundation for the Study of Environment and the Faculty of Architecture and Urbanism at the Sao Paulo University sought FEN collaboration in the establishment of a special course on the analysis and management of urban development to be partly based on teaching materials developed by FEN.

Appendix II: FEN Publications

Research Reports Published by FEN

1. Panjwani, N. 1985. *Citizen organizations and food-energy alternatives in Indian cities.* 74 pp.
   A description of several innovative community development projects dealing with self-sufficiency.
2. Lewis, C. 1985. *The use of dynamic systems analysis to assess the potential for enhanced output in the rural communities of developing countries.* 44 pp.
   A methodological approach to the selection of technologies and improved agricultural practices.
3. Slesser, M. 1985. *Energy systems analysis in food and energy crop production.* 48 pp.
   A methodological approach for appraising resource implications of planning decisions.
4. Silk, D. 1985. *Food-energy nexus research in Canada.* 55 pp.
   A state-of-the-art review of recent activities in this field.
5. Moulik, T.K. 1985. *Integrated food-energy systems in India.* 34 pp.
   A description of numerous projects involving different forms of bio-energy.
6. Sokona, Y. et al. 1985. *Femme-énergie-alimentation en Afrique de l'Quest.* 60 pp.
   Three case studies of the role of women in providing food and energy in Senegal.
7. Parikh, J.K. 1985. *Farm gate to food plate.* 60 pp.
   An analysis of energy use in the post-harvest food systems of India, Pakistan, Burma, and Sri Lanka.
8. Dayal, M. 1985. *Food-energy nexus activities in India.* 64 pp.
   A description of numerous integrated rural development programmes.
9. El-Issawy, I. 1985. *Subsidization of food products in Egypt.* 74 pp.
   An assessment of food subsidy programmes with special reference to the urban poor.
10. Yeung, Y.M. 1985. *Urban agriculture in Asia.* 44 pp.
    A review of urban agriculture in Shanghai, Lae, Penang, Hong Kong, Singapore, and Manila.
11. La Rovere, E.L., and M.T. Tolmasquim. 1986. *Integrated food-energy systems in Brazil.* 66 pp.

A description of on-going experiments involving the integrated production of food and energy.

12. Finquelievich, S. 1986. *Food and energy in Latin America: Provisioning the urban poor.* 51 pp.
    A description of state and community-based initiatives designed to increase the availability of food and energy to low-income people.

13. La Rovere, E.L. 1986. *Food and energy in Rio de Janeiro: Provisioning the poor.* 59 pp.
    An analysis of different approaches (ranging from community-based to government-planned) to improving access to food and energy by low-income people.

14. Sanyal, B. 1986. *Urban cultivation in East Africa: People's response to urban poverty.* 75 pp.
    A socio-economic analysis of urban agriculture activities with special reference to Lusaka, Zambia.

15. Parisot, R. 1986. *Environmental impacts of food and energy production in India.* 42 pp.
    An analytical review of recent publications and on-going projects and programmes.

16. Perez, M.R., 1986. *Sustainable food and energy production in the Spanish dehesa.* 53 pp.
    An ecological analysis of traditional agricultural activity in open, savanna-like woodlands.

17. Wisner, B. 1986. *Food-energy nexus and ecosystem.* 56 pp.
    An analytical report on the international symposium of the same name held in New Delhi.

18. Khouri-Dagher, N. 1987. *Food and energy in Cairo: Provisioning the poor.* 64 pp.
    A critical analysis of government policies on food and energy from the household perspective.

19. Pincetl, S. 1987. *The food-energy nexus in California.* 60 pp.
    An analysis of how the agriculture industry responded to the energy crisis of 1973–1974.

20. Pinton, F. 1987. *Systèmes alimentaries en forêt colombienne.* 50 pp.
    An analysis of the use of energy in two food systems found in the tropical forest of Colombia.

21. Peemans, J.-P. 1987. *Social impact of food and energy technologies.* 62 pp.
    An evaluation of the evolution of dominant food and energy systems and of the perspective for alternative developments in this field.

22. Wade, I. 1987. *Food self-reliance in third world cities.* 46 pp.
    An examination of the potential for improving community-based food production, including a comparative analysis of such activities in Manila, Lusaka, and Mexico City.

23. Tricaud, P.-M. 1987. *Urban agriculture in Ibadan and Freetown.* 45 pp.
    A description of the extent, interest and possible actions to develop urban agriculture in the capitals of Oyo State, Nigeria, and Sierra Leone.

24. Supachat, S. 1988. *Access to food and energy in Thailand.* 58 pp.
    A socio-economic description of the background and current trends in both rural and urban areas.

25. Diallo, S., and Y. Coulibaly. 1988. *Les déchets urbains en milieu démuni à Bamako.*

47 pp.; Ba, B. 1988. *Alimentation-énergie en millieu urbain démuni: le cas de Sénégalis-Dakar*. 34 pp.
Case studies on urban wastes and access to food and energy by the urban poor in two West African cities.

26. Kleer, J., and A. Wos, eds. 1988. *Small-scale food production in Polish urban agriculture*. 68 pp.
The results of an interdisciplinary research project on the history, development, and current status of allotment gardens in Poland.

27. Sachs, I., and D. Silk. 1988. *Final report: 1983–1987*. 38 pp.
A description of activities undertaken by the Food-Energy Nexus Programme, including lists of all publications, projects, and meetings.

Occasional Papers and Co-publications of FEN

Bobo, L. 1984. *Energie et alimentation au Sénégal: comparaison de deux systèmes de pêche*. CIRED/AFME, Paris. 108 pp.

Butin, V. 1984. *De la forêt à l'irrigation: le cas du gazogène en Inde*. CIRED/AFME, Paris. 120 pp.

Catao Aguiar, S., and J. Jose Farias. 1986. "PRONATURE: Brazilian solution for industrial wastes producing energy and food." Occasional paper no. 1/86. 24 pp.

Cordova-Novion, C., and C. Sachs. 1987. *Urban self-reliance directory*. IFDA/UNU, Paris. 214 pp.

Gutman, P., and G. Gutman. 1986. *Agricultura urbana y periurbana en el Gran Buenos Aires, experiencias y perspectivas*. CEUR/UNESCO, Buenos Aires. 275 pp.

Gutman, P., G Gutman, and G. Dascal. 1987. *El campo en la ciudad, la produccion agricola en el Gran Buenos Aires*. CEUR/UNESCO, Buenos Aires.

Legal, P.-Y. 1984. *Problèmes du développement rural au Cap Vert*. CIRED/AFME, Paris. 87 pp.

Panwalker, V.G. 1986. "Food-energy nexus experiments in Bombay and New Bombay (final report)." Occasional paper no. 2/86. 23 pp.

Schilizzi, S. 1984. *Interface alimentation-énergie/The food-energy nexus: Bibliographie classée*. CIRED/AFME, Paris. 45 pp.

SEMPLA. 1985. "*América Latina: crise nas metropoles*." Proceedings of the 1984 Sao Paulo seminar. PMSP/CEPAL, Sao Paulo. 143 pp.

Tricaud, P.-M. 1988. *Agriculture urbaine à Freetown et Ibadan*. Ministère des Affaires Etrangers/Organisation et Environnement, Paris. 53 pp.

FEN Research in Other Publications

Deelstra, T. 1987. Urban agriculture and the metabolism of cities. *Food and Nutrition Bulletin*, 9 (2): 5–7.

FINEP. 1985. "Ecossistemas, alimentos e energia." Proceedings of the 1984 Brasilia seminar. Vols. 1–3. UNESCO, Paris. 217 pp. (Primarily in Portuguese with some English.)

————. 1985. "Ecosystems, food and energy." *Proceedings of the 1984 Brasilia seminar.* Vol. 1 (Synthesis report). UNESCO, Paris. 211 pp.

Finquelievich, S. 1985. "Les villes latino-américaines: énergie et alimentation." *Economie et Humanisme,* 282: 28–33.

Gutman, P. 1985. "A la recherche de réponses nouvelles face aux besoins alimentaires et énergetiques dans les villes: le cas argentin." *Economie et Humanisme,* 282: 8–15.

————. 1987. "Urban agriculture: the potential and limitations of an urban self-reliance strategy." *Food and Nutrition Bulletin,* 9 (2): 37–42.

————. 1987. "Pobreza urbana: explorando algunas microsoluciones para macroproblemas." *Desarrollo Economico,* 27 (106): 279–289.

Khouri-Dagher, N. 1985. "Survivre au Caire: l'accès aux aliments." *Economie et Humanisme,* 282: 16–24.

Kleer, J. 1987. "Small-scale agricultural production in urban areas in Poland." *Food and Nutrition Bulletin,* 9 (2): 24–28.

Panwalker, V.G. 1985. "L'agriculture urbaine à Bombay: quelques observations." *Economie et Humanisme,* 282: 25–27.

RECEM. 1986. *Rede de comunicaçao de experiências municipais.* Municipios em busca de soluçoes, Boletin No. 1, CEPAM, Sao Paulo.

Sachs, C. 1985. "Les métropoles latino-américaines face à la crise: expériences et politiques." *Economie et Humanisme,* 282: 34–38.

Sachs, I. 1985. "Manger cult chaque jour." *Economie et Humanisme,* 282: 5–7.

————. 1986. "Social innovations in the urban setting: Scope and evaluation criteria." *IFDA Dossier,* 53, May/June: 42–45.

————. 1986. "Market, non-market, quasi-market and the 'real' economy." Paper presented at the 8th IEA World Congress, New Delhi. In K.J. Arrow, ed., *The balance between industry and agriculture in economic development,* pp. 218–231. MacMillan Press, Basingstoke, U.K., 1988.

————. 1986. "A interface alimentos-energia: na busca de soluçoes locais para problemas globais." In *Ecodesenvolvimento Crescer Sem Destruir.* Terra dos Homens, Vertice, Sao Paulo. 207 pp.

————. 1986. "Work, food and energy in urban development." *Development,* 4: 2–11.

————. 1986. "Facing the crisis in large cities: Work, food and energy in urban development." *Humanokologie und Geographie* (Geographisches Institut, Zurich). pp. 137–159.

————. 1986. "Les grandes villes face à la crise: travail, nourriture et énergie dans l'ecodéveloppement urbain." *Amérique Latine,* 18.

————. 1986. "Trabalho, alimentacao e energia no ecodesenvolvimento urbano." In *Espacos, Tempos e Estrategias do Desenvolvimento.* Vertice Sul, Sao Paulo, 224 pp.

————. 1986. "Food + energy = development." *Work in Progress,* 10(1) (UNU, Tokyo).

————. 1987. "Les champs de l'innovation sociale." *Problèmes Politiques et Sociaux,* no. 572, November. La Documentation Française, Paris.

Sachs, I., and D. Silk. 1987. "Urban agriculture and self-reliance." *Food and Nutrition Bulletin,* 9 (2): 2–4.

Sachs, I., D. Maimon, and M.T. Tolmasquim. 1987. "The social and ecological impact of 'pro-alcohol'." *IDS Bulletin,* 18 (1): 39–45.

Sanchez, V. 1988. "Estructuras de lo cotidiano y funcionamiento de la 'economia real' en las ciudades: mas alla de la dicotomia formal-informal." *Revista Interamericana de Planificion*, XXII(85).

Streiffeler, F. 1987. "Improving urban agriculture in Africa: A social perspective." *Food and Nutrition Bulletin*, 9 (2): 8–13.

Wade, I. 1987. "Community food production in cities of the developing nations." *Food and Nutrition Bulletin*, 9 (2): 29–36.

Yeung, M. 1987. "Examples of urban agriculture in Asia." *Food and Nutrition Bulletin*, 9(2): 14–23.

———. 1988. "Agricultural land use in Asian cities." *Land Use Planning*, 5 (1): 79–82.

Unpublished FEN Reports or Manuscripts

Abdel-Fadil, M. 1987. "Everyday structures and the working of the real economy in greater Cairo." 31 pp.

Asaduzzaman, M. 1988. "Access of urban poor to food and energy in Bangladesh."

Bailey, L. 1986. "Nairobi case study: Public policies leading to greater equity, efficiency and sustainability in urban energy management." 95 pp.

Carmona, L.S. 1987. "Research project on food, energy and public services self-reliance in Mexico City metropolitan area." 46 pp.

CEDAC. 1985. "Energy and food production in Osasco, Sao Paulo, Brazil." 24 pp.

CEPAM. 1987. "Alternative solutions to food production and distribution." Foundation Mayor Faria Lima, Sao Paulo.

CIEDA (Centro de Investigaciones en Energia, Desarrollo y Ambiente). 1987. "Energia y alimentacion en areas urbanas: un estudio de caso en Caracas." 191 pp.

Galilea, S., R. Jordan, and J. Weinstein. 1987. "The real economy of the metropolitan area of Santiago: Beyond the formal/informal dichotomy." 81 pp.

Ganapathy, R.S., and G. Padmanabhan. 1986. "Energy and urban development: the case of Ahmedabad." 145 pp.

GEIC. 1988. "Case studies of integrated food-energy systems in China."

Giner De Los Rios, F. 1986. "Everyday structures and the working of the real economy: Going beyond the formal/informal dichotomy." 86 pp.

Khosla, A. 1987. "The real economy of Delhi." 29 pp.

Leveque, F. 1986. "Recuperacion de los desechos urbanos de Managua, reporte de mision FEN/UNU." 49 pp. (In Spanish; English summary available.)

La Rovere, E. L., and T.K. Moulik. 1987. "Establishing a permanent international network on biomass-based agro-industrial-energy systems: Towards an alternative rural industrialization strategy." 17 pp.

Perlman, J. E. 1987. "Beggars and limousines: Everyday structures and the working of the real economy in New York City."

Robinson, J., ed. 1986. "The United Nations University Food-Energy (Modelling) Project." 347 pp.

Scartezzini, R. 1987. "Everyday structures and the working of the real economy in the city: The example of Rome." 35 pp.

Silk, D. 1988. "Food-energy nexus in Rwanda."

Urquiza, A., and R. O'Ryan. 1987. "Energia domestica en Chile urbano: estudio de un area de bajos ingresos." PRIEN (Programa de Investigaciones en Energia), Santiago. 75 pp.

Notes

Chapter 3

1. Subsequent parts of this chapter borrow heavily from two reports prepared by Dr. Ben Wisner: "Report of the Seminar" included in the proceedings of the 1984 Brasilia seminar (FINEP/UNESCO 1986) and his report of the 1986 New Delhi conference (Wisner 1986).
2. This section is based on the 1987 feasibility report on "Biomass-based Agro-Industrial-Energy Systems" prepared by Drs. E.L. La Rovere and T.K. Moulik for the UNU.

Chapter 4

1. This section is based on the 1987 report, "Communication and Urban Self Reliance", prepared for FEN by Dr. Yves Cabannes of GRET (Technological Research and Exchange Group), Paris.
2. See "Rede de Comunicaçao Experiências Municipais," *Municipios em Buscade Soluções* 1, no. 1 (1986).
3. "La ville, un abîme inconnu, est (vue de loin) une loterie; là peut-être on aura des chances, tout au moins la misère plus libre."

Chapter 6

1. *The Urban Edge* 12, no. 5 (June 1988).
2. *Senhor*, 28 April 1988.
3. *The Urban Edge* 11, no. 6 (July 1987).
4. *ILO Information* 24, no. 1 (1988).

References

Abdalla, I.S. 1979. "Dépaysanisation ou développement rural? Un choix lourd de conséquences." *IFDA Dossier* 9 (July), Nyon, France.

Alburquerque, C., 1986. "Agro-energy and rural development." In *Proceedings of the International Seminar on Ecosystems, Food and Energy*, vol. II. *See* FINEP/UNESCO 1986.

Bairoch, P. 1983. "Tendances et caracteristiques de l'urbanisation du Tiers Monde d'avant hier à après demain (1900–2025)." *Revue de Tiers Monde*, 24(94): 325–348.

Bendavid-Val, A. 1987. *More with less: MEREC*. Bureau for Science and Technology, UASID, Washington.

Boyden, S. 1984. "Integrated studies of cities considered as ecological systems." In F. Di Castri, F.W.G. Baker, and N. Hadley, eds., *Ecology in practice*, vol. II. Tycooly, Dublin.

Braudel, F. 1979. *Civilisation matérielle économie et capitalisme, XVe et XVIIIe siècle, I: Les structures du quotidien: Le possible et l'impossible*. Armand Colin, Paris.

Brown, L., et al. 1988. *State of the world 1988*. W.W. Norton, New York.

Brownrigg, L. 1985. *Home gardening in international development: What the literature shows*. League for International Food Education, Washington.

Brown Weiss, E. 1989. *In fairness to future generations: International law, common patrimony, and intergenerational equity*. Transnational, Ardsley-on-Hudson/United Nations University, Tokyo.

Cardoso, R. 1985. *Energy and food production in Osasco, Sao Paulo, Brazil*. Unpublished report, UNU/FEN, Paris.

Cecelski, E. 1987. "Energy and rural women's work: Crisis, response and policy alternatives." *International Labour Review*, 126(1).

Chakravarty, S. 1987. *Development planning – the Indian experience*. Oxford University Press, New Delhi.

———. 1987. "Development strategies in the Asian countries," Paper delivered at seminar on Alternative Development Strategies, OECD Development Centre, Paris.

Chambers, R. 1983. *Rural development: Putting the last first*. Longman, London.

Cordova-Novion, C., and C. Sachs. 1987. *Urban self-reliance directory*. IFDA/UNU, Paris.

Council on Enironmental Quality. 1980. *Global 2000 report to the president: Entering the 21st Century*. Washington.

Deelstra, T. 1987. "Urban agriculture and the metabolism of cities." *Food and Nutrition Bulletin*, 9(2): 5–7.

Ducos, C., and P.B. Joly. 1988. *Les biotechnologies*. La Découverte, Paris.

El-Issawy, I. 1985. *Subsidization of food products in Egypt*. UNU/FEN, Paris.

ESCAP. 1985. *Poverty, productivity and participation: Contours of an alternative strategy for poverty eradication*. UN Economic and Social Commission for Asia and the Pacific, Bangkok.

FINEP/UNESCO. 1986. *Proceedings of the International Seminar on Ecosystems, Food and Energy*, vols. I–III. Montevideo.

Finquelievich, S. 1986. *Food and energy in Latin America: Provisioning the urban poor*. UNU/FEN, Paris.

Flavin, C., and A.B. Dunning. 1988. *Building on success: The age of energy efficiency*. Worldwatch Paper, Washington.

Freudenberger, C.D. 1988. "The agricultural agenda for the 21st century." *Kidma*, 38(10):2.

Furedy, C., and D. Ghosh. 1984. "Resource conserving traditions and waste disposal: Calcutta." *Conservation and Recycling*, 8: 2–4.

Furtado, D. 1988. "A crise economica contemporanea." *Revista de Economia Politica*, 8(1).

Glaeser, B., ed. 1987. *The Green Revolution revisited*. Allen and Unwin, London.

Goldemberg, J., et al. 1985. "An end-use oriented global energy strategy." *Annual Review of Energy*, 10.

Gutman, P. 1987. "Urban agriculture: The potential and limitations of an urban self-reliance strategy." *Food and Nutrition Bulletin*, 9(2): 37–42.

Hall, D.O., and R.P. Overend, eds. 1987. *Biomass*. John Wiley and Sons, New York.

Hardoy, J. 1982. "Urban development and planning in Latin America." *Regional Development Dialogue*, 3(2).

Hardoy, J.E., and D.I. Satterthwaite. 1986. "Change in the third world: Are recent trends a useful pointer to the urban future?" *Habitat International*, 10(3): 33–52.

Idriss, J. 1989. "An enviornmental priority." *International Herald Tribune*, 30 January.

Jacobs, J. 1984. *Cities and the wealth of nations: Principles of economic life*. Random House, New York.

Kalecki, M. 1978. *Introdugao a teoria do crescimento em economia socialista*. Prelo Editora, Lisbon.

Khouri-Dagher, N. 1987. *Food and energy in Cairo: Provisioning the poor*. UNU/FEN, Paris.

Kleer, J., and A. Wos, eds. 1987. *Small-scale food production in Polish cities*. UNU/FEN, Paris.

La Rovere, E. 1986. *Food and energy in Rio de Janeiro: Provisioning the poor*. UNU/FEN, Paris.

La Rovere, E., and M.T. Tolmasquim. 1986. *Integrated food-energy systems in Brazil*. UNU/FEN, Paris.

Lipton, N. 1977. *Why poor people stay poor: Urban bias in developing countries*. Templesmith, London.

Meier, R.L. 1974. *Planning for an urban world: The design of resource-conserving cities*. MIT, Cambridge, Mass., USA.

Morris, D. 1982. *Self-reliant cities, energy and the transformation of urban America*. Sierra Books, San Francisco.

———. 1983. "City-states: Laboratories of the 1980's." *Environment*, 25(6).

Moulik, T.K. 1985. *Integrated food-energy systems in India*. UNU/FEN, Paris.

————, ed. 1988. *Food-energy nexus and ecosystem.* UNU/Oxford and IBH, New Delhi.

Munck, L., and F. Rexen. 1985. "Increasing income and employment in rice farming areas – Role of whole plant utilisation and mini rice refineries."In *Impact of science on rice*, IRRI, Los Baños.

Newcombe, K., J.D. Kalina, and A.R. Aston. 1978. "The metabolism of a city: The case of Hong Kong." *Ambio*, VII(1).

Panwalker, V.G. 1986. Food-energy nexus experiments in Bombay and New Bombay. UNU/FEN, Paris.

Parikh, J. 1981. "Energy and agriculture interactions." In K. Parikh and F. Rabar, eds., *Food for all in a sustainable world.* IIASA SR-81-2, Vienna.

Parikh, J.K. 1985. *Farm gate to food plate.* UNU/FEN, Paris.

Peemans, J.-P. 1987. *Social impact of food and energy technologies.* UNU/FEN, Paris.

Perez, M.R. 1986. *Sustainable food and energy production in the Spanish dehesa.* UNU/FEN, Paris.

Richards, P. 1985. *Indigenous agricultural revolution.* Hutchinson, London.

Robinson, J., ed. 1986. *The United Nations University Food-Energy (Modelling) Project.* Unpublished manuscript.

Ruddle, K., and W. Manshard. 1981. *Renewable natural resources and the environoment; Pressing problems in the developing world.* UNU, Tokyo.

Sachs, I. 1980. "Developing in harmony with nature." *Canadian Journal of Development Studies*, 1(1).

————. 1982. "The food-energy nexus subprogramme." Proposal prepared for the UNU, Tokyo.

————. 1984. *Développer les champs de planification.* Université Coopérative Internationale, Paris.

————. 1986. Work, food and energy in urban ecodevolpment. *Development*, no. 4.

————. 1987. "Market, non-market, quasi-market and the 'real' economy." In K.J. Arrow, ed., *The balance between industry and agriculture in economic development*, vol. 1. Macmillan Press, Hong Kong.

————. 1989. *Resources, employment and development financing: Producing without destroying.* RIS Occasional Paper 24, Research and Information System for the Non-Aligned and Other Developing Countries, New Delhi.

Sanchez, V. 1988. "Estructuras de lo cotidiano y funcionamiento de la 'Economia real' en las ciudades: mas alla de la dicotomia formal-informal." *Revista Interamericana de Planificion*, XXII(85).

Sanyal, B. 1986. *Urban cultivation in East Africa.* UNU/FEN, Paris.

Sen, A. 1987. *Hunger and entitlements.* Research for Action series, WIDER, Helsinki.

Seshadri, C.V. 1986. *The sugar-food nexus.* Unpublished manuscript.

Silk, D. 1988. *Valorisation des residus organiques pour le développement rural au Rwanda.* Unpublished manuscript. Caisse Central de Coopération Economique, Paris.

————. 1989. "Mécanismes automatiques pour alimenter un fonds mondial pour l'atmosphère." Unpublished report. Mission Environnement-Développement, Secrétariat d'Etat auprès du Premier ministre chargé de l'Environnement et de la Prévention des risques technologiques et naturels majeurs, Paris.

Soedjatmoko. 1981. "UNCNRSE: First steps towards a broader vision?" *ASSET*, 3(7).

Soemarwoto, O. 1981. "Home gardens in Indonesia." Paper presented at the Fourth Pacific Science Inter-Congress, September 1981, Singapore.

Steinberg, E.B., and J.A. Yager. 1978. *New means of financing international needs*. The Brookings Institute, Washington.

Streiffeler, F. 1987. "Improving urban agriculture in Africa: A social perspective." *Food and Nutrition Bulletin*, 9(2): 8–13.

Tricaud, P.-M. 1987. *Urban agriculture in Ibadan and Freetown*. UNU/FEN, Paris.

Wade, I. 1981. "Fertile cities." *Development Forum* (September).

————. 1987. *Food self-reliance in Third World Cities*. UNU/FEN, Paris.

Ward, B. 1979. *Progress for a small planet*. Maurice Templesmith, London.

Wisner, B. 1986. *Food-energy nexus and ecosystem*. UNU/FEN, Paris.

————. 1988. *Power and need in Africa: Basic human needs and development policies*. Earthscan, London.

World Commission on Environment and Development. 1987. *Our common future*. Oxford University Press, Oxford, UK.

Yeung, Y. 1987. "Examples of urban agriculture in Asia." *Food and Nutrition Bulletin*, 9(2): 14–23.

Zhang, G., Z. Ding, and Z. Xiao, 1986. "Introduction to the new integrated food-energy village scheme in Mazhuang Village of Gushen County in Anhui Province." In *Proceedings of the International Seminar on Ecosystems, Food and Energy. See* FINEP/ UNESCO 1986.